The disappearance of time

This is a book about the philosophy of time, inspired by the writings of the great logician Kurt Gödel (1906–78). It evaluates Gödel's attempt to show that Einstein has not so much explained time as *explained it away.* Unlike recent more technical studies, it concerns itself with the metaphysical implications of the reality of time. The book explores Gödel's published and unpublished thoughts on time and existence, with special reference to related discussions in Aristotle, Kant, and Frege. In the course of this investigation, an attempt is made to shed light on the question of the precise sense in which Gödel believed in the possibility of time travel, on the issue of the significance of time for the foundations of mathematics, as well as for an understanding of the nature of human existence.

The disappearance of time

Kurt Gödel and the Idealistic Tradition in Philosophy

Palle Yourgrau
Brandeis University

The right of the University of Cambridge to print and sell all manner of books was granted by Henry VIII in 1534. The University has printed and published continuously since 1584.

Cambridge University Press

Cambridge
New York   Port Chester   Melbourne   Sydney

If the world is beautiful, and its maker good, clearly he had his eye on the eternal.

Plato, *The Timaeus*

. . . projecting directly from the developed parts of mathematics to the whole world, [Gödel] concludes that it is also beautiful and perfect.

Hao Wang, *Reflections on Kurt Gödel*

Published by the Press Syndicate of the University of Cambridge
The Pitt Building, Trumpington Street, Cambridge CB2 1RP
40 West 20th Street, New York, NY 10011, USA
10 Stamford Road, Oakleigh, Melbourne 3166, Australia

First published 1991

Printed in the United States of America

Library of Congress Cataloging-in-Publication Data

Yourgrau, Palle.
The disappearance of time : Kurt Gödel and the idealistic tradition in philosophy / Palle Yourgrau.
p. cm.
Includes index.
ISBN 0-521-41012-6 (hardcover)
1. Time. 2. Gödel, Kurt – Contributions in concept of time.
I. Title.
BD638.Y68 1991
115′.092–dc20 91-6646
CIP

British Library Cataloguing in Publication Data

Yourgrau, Palle
The disappearance of time : Kurt Gödel and the idealistic tradition in philosophy
1. Time
I. Title
115.092

ISBN 0-521-41012-6 hardback

Contents

Preface

Of the three philosophical essays published by Kurt Gödel during his lifetime, only one is devoted exclusively to the study of time. Indeed, "A Remark about the Relationship between Relativity Theory and Idealistic Philosophy" (Gödel 1949) is, from any point of view, a singular essay. Not only does Gödel announce his discovery of new solutions to the field equations of Einstein's General Theory of Relativity, which include the existence of closed, timelike world lines and permit, in a certain sense, time travel, but he goes on to advance considerations to the effect that his new results support the contention that time itself is merely ideal – that is, a subjective appearance or, more strongly, an illusion. As Gödel sees it, then, Einstein has not so much explained time as explained it away. What, however, can Gödel mean by such enigmatic pronouncements, and what force is it reasonable to attribute to them? Are they isolated and idiosyncratic observations, or do they occupy a natural position in his overall philosophy? Indeed, does Gödel, the logician, have a unified and comprehensive philosophy, or are his remarks on time, together with his better-known philosophy of mathematics, only fragments that fall short of the grand overall scheme of things one has come to expect of philosophers? Since Gödel's death in 1978, answers to these questions have begun to emerge. His *Nachlass* has revealed unpublished philosophical papers, including an extended presentation of "Some Observations about the Relationship between Theory of Relativity and Kantian Philosophy" (Gödel unpublished). In addition, some revealing new studies have appeared that shed light on

Gödel as philosopher – Georg Kreisel, "Kurt Gödel," in *Biographical Memoirs of Fellows of the Royal Society* (1980); Solomon Feferman, "Kurt Gödel: Conviction and Caution" (1984); and, most important, Hao Wang, *Reflections on Kurt Gödel* (1987) – the last constituting, to my knowledge, the only full-length study of Gödel the philosopher in print. [Wang (1974), *From Mathematics to Philosophy,* benefited from contributions from Gödel himself.]

In addition to these more general studies, there have appeared in recent years works addressed specifically to philosophical issues raised by Gödel's contributions to the theory of time, most notably Howard Stein, "On the Paradoxical Time-Structures of Kurt Gödel" (1970) and "Introductory Note to K. Gödel (1949)" (1990), and Paul Horwich, "Time Travel" in *Asymmetries in Time* (1981). Finally, the first two volumes of Gödel's *Collected Works* (1986, 1990) have now appeared, containing all of his previously published papers. It would seem, then, that the time is ripe for a more extended treatment of the philosophical ramifications of Gödel's writings on time. I have already, in a "Review Essay: Hao Wang, *Reflections on Kurt Gödel*" (1989), attempted to put into perspective Professor Wang's philosophical portrait of Gödel, emphasizing aspects of Gödel's thought that Wang had not focused on – for example, the close relationship of Gödel's mathematical Platonism to Frege's and the structural similarity of Gödel's thoughts on mathematics, time, and theology with respect to the thesis that the possible has strong implications for the actual. And in a volume on *Demonstratives* (1990) that I have edited, I have reproduced Gödel (1949) in a setting that will, I hope, help to put Gödel's elusive thoughts into context.

This monograph is devoted to a philosophical consideration of Gödel's papers on Einstein and idealistic philosophy, and to an examination of various metaphysical questions concerning time to which his work gives rise. As will be seen, no attempt is made here to evaluate the purely technical side of

Gödel's results. Much has been written, and continues to be written (see, for example, David Malament 1984 and Lawrence Sklar 1984), on the formal aspects of Gödel's ideas – that is, on those aspects that relate most directly to questions of physics or mathematics. I am, by training and inclination, not the one to add to that discussion. I believe, however, that something is missing from discussions of the dramatic philosophical conclusions that Gödel was at pains to draw from his results. What follows is intended to help fill this in. Logicians and philosophers of space–time physics may well be taken aback by the introduction, in the pages that follow, of metaphysical issues associated with Parmenides, Kant, McTaggart, and others, concerning reality, idealism, becoming, and so on, but I make no apologies for this, since it is Gödel himself who has brought up these ideas. To shy away from metaphysical questions that Gödel himself raises would betray an inclination not to take him seriously as a philosopher.

In the completion of this project, I have greatly benefited from discussions over a period of years with Hao Wang, who has shared with me some of the fruits of his long association with, and reflections on, Gödel. In particular, his remarks have helped me to gain some insight into Gödel's philosophical worldview. I have also benefited from an extended correspondence with Michael J. White over issues related to the logic of time, with special reference to Aristotle and temporal indeterminism, as well as from his detailed comments on earlier versions of this manuscript. And I would like to thank Charles Parsons for his helpful comments on Chapter 6, Robert Tragesser and James Van Cleve for valuable suggestions about the manuscript, David Pitt for preparing the index, and an anonymous referee for Cambridge University Press for a number of helpful suggestions and comments. Finally, I wish to thank Elliott Shore, librarian at the Institute for Advanced Study, for granting me permission to quote from Gödel (unpublished), and the Special Collections Department of Princeton University's Firestone Library for providing me with a reproduction of the drafts of Gödel's papers, which I have done my best to assemble into a working manuscript. These drafts have proved especially illuminating, since by Gödel's indications it appears that much of the most interesting philosophical material may have been destined by him to be eliminated in the final draft. One must therefore bear in mind, when I provide excerpts from these unpublished manuscripts in the pages ahead, that the material quoted may well – for whatever reason – not have been intended by Gödel for publication. Hao Wang has reassured me that the inclusion, nevertheless, of this material would not represent a violation of Gödel's wishes on such matters.

At various points in the book I have drawn on previous works of mine published elsewhere. Specifically, for pp. 83–7, Yourgrau (1985c); for pp. 99–102, Yourgrau (1986); for pp. 121–2, Yourgrau (1987c); and for pp. 124–6, 147–8, and 151–2, Yourgrau (1987b).

1

The reception of Gödel's results

> I would like to say first that I *don't* consider my work a "facet of the intellectual atmosphere of the early 20th century," but rather the opposite.
>
> Kurt Gödel, [Hao Wang, *Reflections on Kurt Gödel,* 1987, p. 20]

In the 1940s, Kurt Gödel, having already achieved striking and fundamental results in the foundations of mathematics, turned his attention to the universe. The occasion for the publication of these new reflections was an invitation to honor Albert Einstein, whose own work, decades before, had ushered in the era of relativistic cosmology. The result of this auspicious occasion was the beautiful and enigmatic "A Remark about the Relationship between Theory of Relativity and Idealistic Philosophy" (Gödel 1949). An even more suggestive essay survives, unpublished, in his *Nachlass,* "Some Observations about the Relationship between Theory of Relativity and Kantian Philosophy" (Gödel unpublished). In these papers Gödel cast a philosophical eye on the ancient riddle of time, "that mysterious and seemingly self-contradictory being, which, on the other hand, seems to form the basis of the world's and our own existence" (Gödel 1949). By Gödel's lights his new solutions to the equations of general relativity lend themselves to a philosophical interpretation that is in a strong sense "idealistic," providing support from an unexpected quarter to those thinkers, from Parmenides to Kant and McTaggart, who had all along been skeptical of the ultimate significance of time.

Now it has been some time since Gödel left us, but I cannot help feeling that the riddle of his enigmatic remarks has yet to be resolved. Indeed, the very first commentator on Gödel (1949), namely, Einstein himself (Einstein 1949), simply brushed aside the second half of Gödel's title ("Entirely aside from the relation of the theory of relativity to idealistic philosophy . . ."). In fact, we read in Wang (1987) that Einstein tended to find such interest in the relationship of Kant to relativity theory rather amusing: "Kant's much praised view on time reminds me of Anderson's tale of the emperor's new clothes, only that the form of intuition takes the place of the emperor's clothes!" (p. 74). Other commentators are divided on the question of what to make of Gödel's assertion that in the so-called rotating universes he discovered (the R-universes, in Gödel unpublished) that "it is possible . . . to travel into any region of the past, present, and future, and back again, exactly as it is possible in other worlds to travel to distant parts of space." Čapek (1961) and Popper (1982), for example, chide Gödel for entertaining the senseless idea that "time travel" could really be possible, Čapek attributing Gödel's mistake to his excessive "spatialization" of time (i.e., likening time to a "geometrical space"). Howard Stein (1970), David Malament (1984), Lawrence Sklar (1984), and Paul Horwich (1987), on the other hand, take Gödel's results to have serious implications for the theory of time, and with considerable care investigate the formal and philosophical ramifications of these results. As Malament puts it: "We do have good evidence that the Gödel model does not describe our universe. I am only proposing here that it cannot be ruled out *a priori*. Neither can any other cosmological model just because it allows for the possibility for time travel" (p. 99).

Now what I find disturbing about the current state of this debate is that the primary philosophical point of Gödel's (1949) paper seems to have been lost or downplayed – namely, that the deduction, from the General Theory of Relativity (or GTR), of the possibility of time travel shows that time is ideal – a mere appearance imposed on reality by a kind of "optical refraction" of the subjective lens of the mind but reflecting nothing objective in the nature of things. Of course, philosophers could hardly have failed to acknowledge that Gödel's purpose in (1949) was to defend an idealistic view of time. What is puzzling, however, is that some writers will, in the same breath, continue to reaffirm (with approbation or without it) Gödel's commitment to *time* travel. But one cannot have it both ways. Yet Horwich (1987), a forceful defender of the possibility of time travel, sees Gödel as his precursor: "It has been claimed by Kurt Gödel (1949) that time travel is in some sense physically possible . . . my aim in this chapter is to defend Gödel's claim" (p. 111). Stein (1970), by contrast, shows a more subtle appreciation of the logic and substance of Gödel's position and makes it clear that "[t]he paradoxical character of the time-structure of his model is emphasized by Gödel in (1949), where his chief aim *is to put in question* our ordinary notions about time" (p. 591). Stein himself, however, is less impressed than Gödel is by the paradoxical implications of time travel, and after asking us to "[n]ote Gödel's . . . implied acknowledgement that 'actual feasibility' of time-travel *would* constitute grounds for rejection of such structures a priori," he goes on to suggest that "[i]n my view, this concession is not warranted" (p. 594).

What must be admitted, of course, is that Gödel believes he has shown the compatibility with the GTR of universes permitting time travel – in Stein's words, a situation in which, "[i]f "decisions" can be taken and "free actions" performed in such a universe – and if adequate energy-sources are available – then a person in such a world can decide, at any point in his career, to visit the space–time vicinity of any epoch of his own past." (1970, p. 591). But it is this very fact that Gödel takes to indicate that *t*, the standard variable for time, should not be read here as standing for genuine, successive *time*. But if there is no genuine time, there can be no genuine time travel. This somewhat confusing situation is in certain respects comparable to a similar difficulty in the interpretation of Berkeley – an idealist in a different (though related) sense. Berkeley was always at pains to insist that his idealism about the physical world did not commit him to the view that ordinary physical objects like tables and chairs fail to exist. He insisted that chairs really do exist but went on to remind us that by his philosophy this does not imply that matter outside the mind exists, since he wishes us to interpret all talk of tables and chairs and other physical objects in terms of mental ideas. Similarly, Gödel describes the R-universe as permitting time travel, but only if we do not read "time" here as designating genuine, successive time, but rather as denoting a relativistic formal simulacrum of the real thing.

We must be careful, however, to distinguish Gödel's idealism from Berkeley's in the following two respects. First, Gödel means only to assert that time is an "appearance" or illusion, not that it is mental. (Indeed, if the mental realm admitted genuine temporal succession, Gödel would surely have concluded not that time as such is an illusion, but only that its sphere of application is limited.) Second, Gödel's idealism applies only to time, not to space or to space–time. [Indeed, an adequate explication of Gödel (1949) must explain this fact. We will come to it later.]

It is easy to see, then, how Gödel's deduction of the possibility of time travel could come to overshadow his use of his results to conclude the ideality of time. The reasons for this reversal of emphasis, however, in my opinion, run deeper. Gödel has always been sparing of the details of his overall philosophy, which, moreover, he has stressed, goes very much against the temper of the times. Indeed, only now have we begun to see the extent to which Gödel suppressed in his (already very sparse) published writings the heuristic philosophical background against which he himself believed it necessary to place his formal results in order for their true implications to be realized. Already in *From Mathematics to Philosophy* (1974, p. 9), Wang had reported how Gödel had even suppressed, in his published exposition of his early Incompleteness Theorems, that "the heuristic principle of my construction of undecidable number theoretical propositions in the formal systems of mathematics is the highly transfinite concept of 'objective mathematical truth' as *opposed* to that of demonstrability" [whereas the printed text of (1930) used only the notions of consistency and ω-consistency]. It is clear that Gödel feared that the philosophical climate, under the large shadow cast by Hilbert, in which the concepts of truth and of the transfinite were held in suspicion, made it advisable for a prudent man to isolate his formal results from such philosophical contamination.

More recently, Solomon Feferman, in "Kurt Gödel: Conviction and Caution" (1984), has mused on "the tight rein [Gödel] placed on the expression of his true thoughts" (p. 559). After remarking the fact mentioned previously, Feferman points out, in addition, that "Gödel never did state publicly (until his communications to Wang reported in 1974) what he thought of the views of the Vienna Circle" (p. 560), although the evidence shows that "it is safe to say that Gödel already had firm general platonistic views by the time he came in contact with the Vienna Circle" (p. 560). Feferman even speculates that Gödel's excessive caution in regard to (what he considered) the dominant "blindness (or prejudice), or whatever you may call it" of the times may have kept him from going ahead with fundamental research, for example, on computable (λ-definable, general recursive) functions (as Turing did), or on truth (as Tarski did). [This last claim, however, raises an issue I cannot go into here, namely, Gödel's anticipation of Tarski's results on the undefinability of truth. See Wang (1987), pp. 85, 273.] "Gödel's spreading fame," says Feferman, "should have reassured him that he would not be laughed off-stage if he were to go beyond the purely (logico-) mathematical formulation of his results" (p. 559).

The striking picture that emerges from Feferman's essay is

reaffirmed by Wang's revealing new study, *Reflections on Kurt Gödel* (1987). Gödel's quarrel with the zeitgeist, it turns out, was far more considerable than even Feferman had suggested. Strongly influenced by Plato and Leibniz (and, later, by Husserl), Gödel had devised his own "ontological argument" for the existence of God ("my belief is *theistic,* not *pantheistic,* following Leibniz not Spinoza", Wang 1987, p. 8), disliked the *Tractatus* ("because it purported to show that philosophy is not possible," p. 77), and believed that "there is mind separate from matter" (p. 197). In light of such views, it is perhaps a bit unfair to characterize Gödel's caution as "excessive."

Especially striking is Gödel's resistance to the contemporary bias toward "formalism" (in a broad sense), the tendency to recognize nothing in a domain beyond those features captured and represented in some chosen mathematical formalism. Already in his Incompleteness Theorems, of course, Gödel had provided formal evidence (!) against the most precise cise and well-formulated program of formalism of this century, the so-called Hilbert program, which identified (or replaced) mathematical truth with provability in a formal system [where the latter notion was given (relatively) precise mathematical sense – only fully precise, for Gödel, after the work of Alan Turing on computability]. As Gödel wrote to Wang, "formalists considered formal demonstrability to be an *analysis* of the concept of mathematical truth and, therefore, were of course not in a position to *distinguish* the two" (Wang 1974, p. 10). Gödel's results, and his essays on the philosophy of mathematics (1944 and 1964), suggest that formalisms do not "speak for themselves" as to their full significance, but require something more, something informal or "intuitive," if you wish, for their proper interpretation. A striking contemporary example of extreme formalism (to continue to use the term in a broad sense) is the employment of Hilary Putnam (1985) of the Löwenheim-Skolem Theorem to argue that the formalization of Cantor's proof of the existence of nondenumerable sets subverts Cantor's intentions, since the formal proof can always be reinterpreted to prove the existence of at most a denumerable totality. This effect – the "Skolem Paradox" – is achieved precisely by refusing to put the chosen formal system into a wider interpretive context, thereby, in effect, forcing the formalism to speak for itself. [For a reasonable response to this paradox, with special reference to Putnam (1985), see Paul Benaceraff (1985), who associates his approach explicitly with Gödel's.]

The Skolem Paradox represents an extreme version of mathematical formalism insofar as it combines two salient features of the late-nineteenth- and early-twentieth-century rigorization of mathematics – axiomatization and formalization (i.e., representation in a first-order formal system in the technical sense). In a broader sense, however, the attempt to capture the essential features of a domain by purely mathematical–structural means (at the expense, for example, of the qualitative or intuitive) represents the fundamental idea of formalism. Most important for our purposes is the mathematization (in the broad sense) of nature in the historical development of natural philosophy, especially physics, from Descartes's geometrical method (and reduction of matter to "pure extension") and arithmetization of geometry to the geometrization of time itself in Einstein's Special Theory of Relativity (STR) and GTR. In fact, the historical (as opposed to the mythical) Descartes was not so pure in his geometrization and arithmetization as in the modern development culminating in Einstein. As David Lachterman has argued, "for Descartes graphic constructions [of geometric figures] *are* the outward counterparts and sensible signs of the inward activity of mind" (1989, p. 148). Not so, surely, for David Hilbert (1902). And not so for Einstein. Indeed, "[t]he primary object of Einstein's profound researches on the forces of nature," writes G. J. Whitrow, "has been well epitomized in the slogan, 'the geometrization of physics,' *time being com-*

pletely absorbed into the geometry of a hyper-space" (1980), p. 4 [my emphasis]. To what, however, can we attribute Gödel's interest in the geometrization of time?

We have already mentioned the official occasion for Gödel's reflections on Einstein's treatment of time, but it should not be thought that his concern with this subject was merely fortuitous. Wang (1987) affirms that Gödel attributed his interest in time to his early work on Kant and Leibniz. Perhaps, then, we can offer the following as a kind of "rational reconstruction" of Gödel's interest in time, whether or not Gödel himself consciously saw himself this way. [And as Imre Lakatos has observed (1987, p. 4), if such rational reconstructions strike some as mere caricatures of history, "one might equally well say that both history and the way things actually did happen are just caricatures of the rational reconstruction" – a point of view that would surely have appealed to the archrationalist Gödel.] All along, then, Gödel viewed his well-known mathematical-logical work as the investigation of an objective "platonic" domain that does not depend on the empirical. "[L]ogic and mathematics," he wrote "(just as physics) are built up on axioms with a real content which cannot be explained away" (Gödel 1944, p. 132). It was natural, therefore, for Gödel (as Frege had done before him) to make a sharp distinction between epistemology and ontology, thought and object of thought, proof and truth (distinctions that must be more blurred for other philosophers of mathematics, such as Hilbert, Brouwer, and Wittgenstein, who reject such platonism).

With Plato, moreover, Gödel also shares the recognition of another crucial distinction, that between the temporal and the timeless or "eternal" (in Plato's sense), where the realm of pure mathematics is confined to the latter, whereas the former applies only to the empirical-"sensible" realm; for, as Whitrow has written, "although Plato regarded time as an essential feature of the sensible world, he rigorously excluded it from the science of pure geometry which he associated with, and only with, the eternal world of ideal forms" (1980, p. 179). Gödel, with Plato, also regarded the world of time as somehow "ontologically suspect," as I will argue in due course. With Plato, finally, Gödel, I suggest, attempted to integrate the two realms, the eternal and the temporal, to "save the phenomena," though Gödel's final result turned out to be far more negative than Plato's. A crucial reason, clearly, for Gödel's negative result (his failure, as it were, to save the phenomena) is that he faced the problem of integration in the context of the extreme mathematization of the temporal represented by Einstein's Relativity Theory.

To be clear, now, on the extreme sense in which Relativity Theory represents a mathematization of time, we should recall earlier systematizations by Aristotle and Newton (developed by Kant). Aristotle and Newton had, of course, applied mathematics to time, Aristotle, famously, defining time itself as "the number [i.e., the numerable aspect] of motion with respect to before and after," and Newton's discovery of the calculus was especially designed to provide a measure of continuous motion (in time). Yet, for all that, the characteristic "mode of being" of time, namely, its "flux," was considered representable, at best, intuitively, or by analogy, so that the idea of the "geometry" of the one-dimensional time line was clearly seen as, at best, a highly useful (if potentially deceptive) *metaphor.*

Kant, in his philosophical reformulation of Newton, is explicit on just this point: "Time is nothing but the form of inner sense, that is, of the intuition of ourselves and of our inner state . . . and just because this inner intuition yields no shape, *we endeavor to make up for this want by analogies*" (1965, p. B50; my emphasis). As Mary Tiles puts it, "philosophy of time cannot proceed simply by conceptual or linguistic analysis, but must go via intuitions, nonverbal representations . . . rhetorical devices other than those of logic and argument" (1989a, p. 182). For Kant, the central analogy was that of a line continuously being generated: "we cannot obtain for ourselves a representation of time, which is not [itself] an object of outer intuition, except under the image of a line, which we draw, and . . . by this mode of depicting it alone could we know the singleness of its dimension" (1965, p. B156); moreover, "one represents a line as generated through fluxion and thus generated in time" (1902, vol. 14, p. 53).

For Aristotle, too, there is a guiding and irreducible metaphor or analogy for time: "[Aristotle's] operative simile is that of the 'now' and its present-past to a moving object and its trajectory or to a geometrical point and the [partial] line generated by its 'fluxion.' This dynamic model of time leaves the past fixed but continuously supplemented, and the future at least partially indeterminate" (White 1985, p. 22). What Gödel found, however, in Relativity Theory, was an account of time where the mathematical-formal element completely dominated the intuitive or analogical, an account in which the Einstein–Minkowski geometry of time is now no longer only metaphorically a geometry. And as the future was to bring out, once the notion of the geometry of time was taken literally, the door was open to someone like Gödel to introduce the question of a nonstandard geometry of time – much as Einstein had earlier employed a nonstandard geometry for space or space–time.

Precisely those features at the heart of the analogies of time of Aristotle and Kant disappear in Einstein's "geometrization of the temporal": "It appears therefore more natural to think of physical reality as a four-dimensional existence, *instead of, as hitherto, the evolution of a three-dimensional existence.* This *rigid four-dimensional space* of the special theory of relativity is to some extent a four-dimensional analogue of H. A. Lorentz's rigid three-dimensional aether" (Einstein 1961, pp. 150–1; my emphases). Again, Louis de Broglie: "In space-time, everything which for each of us constitutes the past, the present, and the future is given in block, and the entire collection of events, successive for us, which form the existence of a material particle is represented by a line, the world-line of the particle" (Schilpp 1982, p. 114).

Note the difference between the Einstein-Minkowski world line and the continuously generated line of Aristotle and Kant. As Einstein and de Broglie bring out clearly, what Relativity Theory seems to offer us is in fact a "spatialization" of time [to borrow Čapek's term from (1961)]. What constitutes time as "spatial" here, and thus susceptible to a purely geometrical treatment, is, I suggest, what is the essence of any "space" in the most general sense of this term – namely, not extension, but rather *the ontological-neutrality of all the positions defined by such a manifold*. By this I mean that something's *location* in a genuine space [whether it be physical space or the "mathematical space" of the natural number sequence (construed platonistically)] has no effect on the status of its existence. This more general definition of a space also makes better sense, I believe, of Gödel's conception of *sets* as "quasi-satial" (Wang 1987) than the classical notion of a physical space. Gödel presumably meant that a set is a paradigmatic *platonic* object whose members do not come into being successively. This is then to be contrasted with the "temporal mode of being" (construed pre-relativistically), in which to exist *is*, precisely, to exist *at* a designated, or privileged, *temporal location* – namely, *now*, or in the present. To "spatialize time," then, is to collapse the latter mode of being into the former, as Einstein and de Broglie would seem to be doing in the quotes we have cited.

Whether this spatialization of time is indeed forced on us by Relativity Theory, however, has been questioned, as we will see in due course. Gödel, though, did not question it, and by exploiting the mathematization of time to introduce a kind of extreme or "limit case," he believed he had made the implications, vis-à-vis spatialization, of Relativity Theory inescapable. Now the significance of Gödel's having before him, in Einstein's theories, such an advanced mathematization of time can be appreciated if we recognize the surprising unity that underlies some of Gödel's most startling formal–mathematical discoveries. A clue to this unity can be found if we consider a nice distinction introduced by John Myhill (1951). A formal question, let us say, asks us to supply the content of its answer, whereas an informal one demands of us that we supply both form and content. Further, "there is no way of knowing whether an informal question is objective or subjective until the question is answered" (p. 42). Now Gödel, I suggest, has a peculiar genius for transforming previously informal questions into formal ones and then settling the remaining question of content by introducing formal constructions that limit the possible contentful interpretations that they can admit. I call this the introduction of a "limit case," by analogy with powerful limitation results like the Löwenheim–Skolem theorem. This way of looking at things, I believe, allows us to tie together Gödel's dramatic formal–mathematical results concerning Frege, Hilbert, and Einstein, as well as to begin to see why these proofs are of such enduring, if ambiguous, philosophical significance.

In the case of Frege, indeed, Gödel's Completeness Theorem actually provided the reverse of a limitation result, since what he demonstrated was that on the level of the formalization of first-order predicate logic with identity, first presented in Frege's *Begriffschrift,* the formal (or syntactic) property of being provable was coextensive with (and so, formally indiscernible from) the contentful (or semantic) property of being valid (or true under all interpretations). Here, then, proof can indeed simulate truth, and thus one cannot impose a limit on either interpretation of the formalism. In contrast, for Hilbert's ideal, the complete formalization of each branch of mathematics and the replacement of the concept of mathematical truth with that of provability in a suitable formal system (itself to be proved consistent by finitary means), Gödel's result was genuinely limitative. The Incompleteness Theorem showed that here proof could not simulate truth. In particular, Gödel demonstrated an intrinsic limitation on the semantic or contentful interpretation that can be imposed on the syntactic predicates in Hilbert's formal systems. No predicate, *T,* can consistently, in this context, bear the semantic interpretation "is true." In effect, then, Gödel showed, by purely formal means, that here Hilbert could not replace (nonfinitary mathematical) content by paying attention only to form. This spelled the end, then, of Hilbert-style (mathematical) formalism. [To be fair, however, one should add that this orthodox view has been challenged of late. See Detlefsen (1979), (1986), and (1990a).]

These results of Gödel's relied, of course, on the very mathematization of logic and metalogic introduced, respectively, by Frege and Hilbert. Exploiting, next, Einstein's geometrization of time – itself a kind of formalization – Gödel was able to construct a mathematical model of a possible universe in which, by his lights, there are essential constraints on the possible contentful, intuitive interpretations the model can bear. In particular, since in such a universe there are closed, timelike world lines, and, in a certain sense, the possibility of time travel, Gödel concludes from this that the standard interpretation of *t,* here, as denoting *time* – genuine successive time, which represents the objectivity of becoming – is clearly impermissible. This reading of the model relies on the fact that Einstein has provided not just a *time geometry,* but rather a *geometrization* of the temporal. The distinction is crucial. Many domains can be described as occupying a certain logical space, which, in turn, may have a distinctive geometric signature. The geometrization of such a domain, however, consists in the elimination or reduction of all qualitative features in favor of the geometrical or structural.

To put this distinction to work, consider Erwin Schrödinger's geometry of color space (see Moore 1989, pp. 120–9). Schrödinger had no intention of reducing the qualitative properties of colors to the purely formal, structural aspects of the geometry he wished to employ. He introduced a geometry, then, but he did not attempt to geometrize color. In contrast, the Einstein–Minkowski geometry of space–time *is,* I suggest, put forward precisely as a geometrization of time, that is, to replace or reduce the intuitive, qualitative features of time to the geometrical (recall the earlier characterization above by Whitrow). Talk of the intuitive content of temporal intuition – the "given," if you will, of time consciousness – gives way to a purely structural representation of the form of time. The parallel with Hilbert is obvious, as is brought out in the following passage from Hermann Weyl (1949): "Intuitive space and intuitive time are thus hardly the adequate medium in which physics is to construct the external world. No less than the sense qualities must the intuitions of space and time be relinquished as its building material; they must be replaced by a four-dimensional continuum in the abstract arithmetical sense. . . . What remains is ultimately a *symbolic construction* of exactly the same kind as that which Hilbert carries through in mathematics" (p. 113).

Now, clearly, this is also how Gödel reads the Einstein–Minkowski geometrization of time. Nevertheless, he is well aware that there are those who persist in maintaining the compatibility of relativity theory with the interpretation of t as successive, intuitive, unfolding time. With the derivation of the world model of the Gödel universe, however, as a solution to the field equations of General Relativity, Gödel has constructed a limit case for the relativistic geometrization of time. That is, he has produced a formal model that essentially limits the possible intuitive, contentful interpretations it can support (due, as we saw, to such features of the model as the presence in it of closed, timelike world lines). Thus, just as the Incompleteness Theorem demonstrated in regard to the Hilbert program that, in that context, mathematical truth cannot be simulated by formal proof, so, in regard to Einstein, the construction of a formal–mathematical model of the Gödel universe demonstrated, by Gödel's lights, that in this context, *t* cannot be given the intuitive, contentful interpretation of denoting that successive, unfolding time that issues in objective temporal becoming. (As we will see in detail later, Gödel in fact goes further and constructs a delicate modal argument from the latter result to the effect that even in the actual world *t* cannot denote intuitive, successive time.)

Now a natural question arises at this point as to why Gödel did not reach the same conclusion in regard to Einstein's geometrization of time as he did for Hilbert's attempted formalization of mathematics, namely, that his surprising discoveries demonstrated the limits of the formalism, rather than the disappearance of an intuitive content not just as a possible interpretation of this system but, as it were, absolutely. We will address this issue in due course. What is to the point here, however, is that in his papers on Einstein – the technical and the philosophical – he appears to have brought a classic, and recalcitrant, informal question (here, the question of the objectivity of temporal becoming) to a sudden and dramatic formal resolution, and again by means of the construction of an ingenious limit case that essentially constrains the possible intuitive *content* that can be expressed by this *form*. Since the power of formalization (in this broad sense) is thus, in Gödel's hands, a tool not for the elimination or detraction of pure form (as in Brouwer) or content (as in Hilbert) (to simplify matters considerably), but rather for the illumination of each by the other, Wang's (1987) characterization of Gödel as engaged in a "dialectic of the formal and the intuitive" seems well said indeed.

Now the notion of a limit case, as I have introduced it, can be usefully compared with that of a "limit concept" as this arises in traditional philosophy. The philosopher as such, one might say, attempts to limit or delimit conceptually a complete domain or universe of discourse, which enables him or her to subject this very domain or "whole" itself to the further manipulations or delimitations of pure thought (if perhaps subject now to new rules of manipulation). The mathematician, in turn, follows this line of development with even greater purity or abstractness. Frege, the seminal figure in the mathematization of logic, completed or delimited, in pure thought, the domain of all (quantificational) validities by means of the (first order part of the) formal system of his *Begriffschrift*. This, then, paved the way for Hilbert to introduce the discipline of metalogic, or the formal manipulation of systems of logic as themselves objects of mathematical investigation. Indeed, to take the simplest example, one of the basic steps in mathematics, the move from the finite natural numbers to the single, definite, infinite series that contains all the natural numbers [which Gödel calls the decisive "first infinite jump" (Wang 1987)], is a particularly pure instance of the employment of a limit concept (here, the concept of "*all* natural numbers"). One leap of idealization leads to others. Cantor introduces the pure limit concept of "actual infinity," thus delimiting the infinite series or infinite sets themselves, and proceeds to construct new rules for the manipulation of the corresponding transfinite numbers (cardinal and ordinal) – a transfinite arithmetic by analogy with classical finite arithmetic.

What emerges even from this brief consideration, I believe, is that limit concepts typically involve, implicitly or explicitly, a powerful universal quantification, or notion of "all," that is always in danger, if unchecked, of leading to paradox. Aristotle himself, as we see later, who first systematized logic, and thus tried to capture or delimit the domain of all logical validities, quickly ran into paradox when he considered the application of one of his logical principles, the excluded middle, to another domain given by a limit concept, that of "all future times." And Kant, of course, devoted a good deal of attention in the "Dialectic" of his *Critique of Pure Reason* to such limit concepts as the limit of *all* time, of *all* space, of *all* causal chains, and so on, which, if employed "uncritically" (in his sense), would, he argued, lead reason, itself, into antinomy (see Strawson 1985, p. 34). Clearly, cosmology, which, as a systematic science, treats the entire universe as a complete, delimited, whole, would thus be put very much into question. Actual antinomy, unfortunately, did arise, precisely in the rigorous domain of logic, when Frege, in attempting to derive the basic laws of arithmetic from the basic laws of logic, took the extension of a concept – that is, the totality that includes all things that fall under the concept – to be itself a completed whole, and thus an object. For, by Axiom V in 1967b, if two extensions are identical, then for *all* objects as arguments, the corresponding concepts yield the same values – even when the argument is the extension of the concept, itself. This, of course, leads to Russell's paradox. Indeed, even today we have to take great care with our use of "all" after Cantor's proof that even his powerful set theory cannot contain "the set of all things," since there is no such set. In fact, the very idea of Reality itself can be seen as, so to speak, the very limit of all limit concepts, and so the most liable to lead, if treated 'uncritically,' in Kant's sense, to antinomy [as will appear shortly when we come to discuss Parmenides, one of the thinkers Gödel (1949) cites as having gotten it right about the relationship of time to Reality].

Gödel's construction of limit cases, then, is in the great philosophical–mathematical tradition of employing limit concepts to enable the mind to grasp or delimit the most distant reaches of reality. His formal constructions, though, are two-faced. In themselves, they are theorems in mathematics proper, and, as such, are no more subject to dispute than mathematics itself is. As we have seen, however, the results distinguish themselves by admitting alternative interpretations, both within and outside of mathematics, and it is these alternative readings that force the "philosophical-minded" [to use Gödel's term from (1949)] to make hard choices on questions that would themselves have been considerably less visible under less extreme logical – or geometrical – constructions. In order to see, then, how Gödel has managed to force us to come to a hard decision about the nature of time, let us proceed to those cosmological results with which our discussion began.

2

Gödel's idealism

> You ask me about the idiosyncracies of philosophers? There is their lack of historical sense, their hatred of even the idea of becoming, their Egyptianism. They think they are doing a thing *honor* when they dehistoricize it, *sub specie aeternitatis* – when they make a mummy of it.
>
> Nietzsche, *Twilight of the Idols,* (1968), p. 351

In the simplest terms, the relevant details of Relativity Theory, as Gödel found it, are as follows. (Experts will no doubt wish to introduce complications, but for our purposes, what matters most is that we have in front of us a perspicuous and concise overview of those philosophically significant aspects of the theory that influenced Gödel's construction.) In the STR the relativity of (distant) simultaneity, and the absence of a distinguished or "privileged" reference frame, prevent one from speaking, as in classical physics, of a nonrelative "worldwide instant." The "now," in this new context, has only frame-relative significance. "The four-dimensional continuum," Einstein writes, "is now no longer resolvable objectively into sections, all of which contain simultaneous events; 'now' loses for the spatially extended world its objective meaning" (1961, p. 149). The classical absolute notion of simultaneity, as a two-term, transitive relation, is replaced by a new three-term relation – simultaneous at an inertial frame. Nevertheless, there remains in the STR a measure between events that is frame invariant – namely, the interval. The equation for the interval, $ds^2 = dx^2 + dy^2 + dz^2 - c^2dt^2$, establishes that for events with timelike separation, where

$ds^2 < 0$, their temporal order is invariant, – that is, the one precedes the other in all frames of reference. Since timelike separated events are causally connectable, we can say that STR preserves at least "causal" time (order).

In the GTR, however, the global distribution of matter affects the overall curvature of space–time itself, so that from a cosmological point of view, it may be that certain inertial frames are privileged – namely, "those that follow in their motion the mean motion of matter" (Gödel 1949, p. 263). In all the cosmologies projected from the GTR before Gödel, it was possible to fit these so-called co-moving frames into a single, consistent "cosmic" time. So, from the perspective of the STR and the GTR, although absolute, "equably flowing," Newtonian time has disappeared, one can still preserve, compatible with the equations, and given some luck in the distribution of matter, causal time as well as cosmic time.

Now one could well question whether the latter concepts are sufficiently close to what was meant, pre-relativistically, by "time" to justify the continued use of this term from classical physics, but we will reserve such considerations for later. In the present context, what is to be noted is that Gödel discovered that there is, compatible with the GTR, a nonstandard, "unlucky" universe, the so-called rotating or R-universe, where the "compass of inertia" rotates everywhere relative to matter and where one cannot fit together the co-moving frames into a single, consistent, cosmic time. In fact, he showed that in such worlds there exist nearly closed as well as closed, future directed, timelike curves – that is, world lines that, although everywhere timelike, eventually return near, or to, their starting points. In such worlds one can, in effect, "travel" in time in the same way we can now travel in space, and, in particular, though always heading into our causal future, we eventually arrive at our past or present.

What can one say about the meaning of these results? On one level, these are purely mathematical deductions from the GTR, and hence unquestionable if the deductions are sound.

(Some, including Einstein, have questioned their physical significance, but we will come to this later.) Gödel interpreted his discoveries, however, in a dramatic philosophical fashion, suggesting that (1) in the R-universe, the objective lapse of time is an illusion and (2) even though it is an empirical question whether or not ours is an R-universe, the collapse of the objectivity of genuine temporal succession in a universe differing from ours only in certain non-lawlike features concerning the cosmic distribution of matter and motion shows that in our universe, too, time is merely ideal. In order to see why Gödel drew such striking philosophical conclusions from his formal results, it will be necessary to make an extended trip back to the dawn of philosophy, to the old quarrel that philosophers have had with time.

If we take our lead from Gödel and go as far back as Parmenides, we see that the problem seems to be a perceived incompatibility between the concept of *being,* or *reality,* and the concept of *time* (and so *change* and *motion*). Now the term "reality" is often used rather loosely, sometimes even as a mere substitute for "the world." But to appreciate the power of such concepts for serious ontologists like Parmenides and Plato (and Gödel), it is useful to approach their full force gradually. In the primitive sense, reality, understood as the world, was taken to be the earth. Since the earth is soon seen to be but one part of the universe, however, "the universe" replaces "the earth" as the full generalization of the concept of the totality of the physical world. Yet even today, cosmologists are somewhat loose in their use of this term, speaking of a "Big Bang" physical "origin" of the universe, whereas they should in strictness speak rather of the origin of *the rest* of the universe (i.e., besides the Big Bang). (And a similar criticism could be made of talk of *multiple* universes.) Even this large concept is soon seen, however, not to be large enough, since the universe is, after all, only a large object, whereas reality, or what-is (in the broadest sense) seems to include what-is-the-case with this object (and what it contains – namely, us). In a word, "[t]he world is all that is the case. It is the totality of facts, not of things" (Wittgenstein, 1974, 1, 1.1) The facts, however, might have been otherwise, and so if the actual world is the totality of facts that actually obtain, then surely we can speak also of other, merely possible worlds that might have obtained. But now that we have introduced the modalities of actuality and possibility, how can we resist the generalization of "reality" into the "modal space" containing (in some suitable sense of "contain") *all* the possible worlds (including our own, the privileged or distinguished actual world)?

But we are not done yet. For we have still left out certain facts. *That* our possible world, the actual world (let's name it α), is actual is not a "fact" that is "internal" to α. Even if α contains a fact to the effect that α is actual, this is still merely one of the many *possible* facts that *constitute* α, and the question will still arise: Is *this* possible fact actual? Where, however, can this missing fact be found? If it is a "superfact" about "modal space" that in it α happens to be (the) actual (world), then it remains that since (theological scruples like Leibniz's aside) α might not have been actual, this superfact itself might not have obtained. Thus we seem to have arrived at a vicious infinite regress, and the concept of reality, as *all* that is the case, has eluded us. (See, further, Yourgrau 1986.)

The source of our difficulties is clearly the extreme sense of *totality* required of the concept of reality. We have focused on the problem along the dimension of modality, but there are equally disturbing considerations when we take into account the dimension of time. And this is where Parmenides comes in. He says, in the *Way of Truth* (1957, pp. 35–8): "There is only one Way left to be spoken of, namely that it *is*. . . . Nor was it ever, nor will it be, since it is now all at once. . . . Thus it must either be altogether or not at all. . . . And how could what is be going to be in the future? – For if it came into being, it *is* not; nor *is* it, if it is at some time going to be. Thus becoming is extinguished and perishing is not to be heard of." As I read it, Parmenides's reasoning seems to proceed along the following lines. If reality, or being, is the totality of all there is, of "all that is the case" (as Wittgenstein puts it), then if time is real, and so also change, then when yesterday I said that reality is all there is, I misspoke, since I left out something – namely, what was not then but came to be. But if time is *real,* then I cannot simply modify my notion of reality to include, as well, what has been and what will be (the case), since *the real* (the existent) *in time* is (what is in) *the present.* Indeed, Gödel himself reaffirms this when he speaks of "an objective lapse of time (whose essence is that only the present really exists)" (1949, fn. 4). In a word, if the "is" in "*all* that there *is*" is *tensed,* it must always fail to apply to future times, for, as Richard Jeffrey puts it, "[If time and change are real] Reality *grows* by accretion of facts" (1980, p. 253; my emphasis). Yet if the future is to add to the present, it must draw on, so to speak, what is (now) nothing, which is impossible, since "it is not possible for 'nothing' to be" (Parmenides op. cit., p. 31), "nor shall I let thee say or think that it came from what is not" (op. cit., p. 36). But if it is *tenseless,* we have already eliminated time as real, or objective, and the Parmenidean tension between time and reality has been "resolved" only by dispensing with the former. (I am taking tense here as the semantic correlate of the "objective lapse of time.")

The only hope for the reality of time, then, would be to try, again, the first option. But at this point Parmenides would surely object: If reality is no illusion, it can hardly 'grow' (as Jeffrey suggests) or get bigger! And what of Wittgenstein's "*all* that is the case?" If this is a *serious* "all" (as it is surely meant to be), how could it expand to include new facts? "[A]n expanding universe is one thing, an 'accumulation of Reality' (and thus of truth) another" (Yourgrau 1987b, fn. 20). Yet Gödel himself seems to represent reality as expanding if time is real when he says that "[t]he existence of an objective lapse of time, however, means (or at least is equivalent to the fact) that *reality consists of* an infinity of layers of 'the now' which *come into existence successively*" (1949, p. 262; my emphasis). If he meant only that reality already contains all the "nows," he would not have added that these come into existence successively. But if what reality consists of comes into being successively, then reality itself must come into existence successively (if time is real).

One should recall, at this point, Gödel's methodological employment of limit cases in his formal work and our earlier remarks on the concept of reality as the limit of all limit concepts. Kant's warning concerning the antinomies that may attend the 'uncritical' employment of such concepts should now, I think, appear to be justified. Still, if we are to engage in ontology at all, it is hard to see how *reality* can be an illusion, and so, it might seem, *time* must be, as Parmenides and Gödel would have it. Indeed, perhaps we can now understand an otherwise mysterious remark in Gödel (unpublished), where he suggests that his cosmological results only confirmed his suspicions about the legitimacy of the concepts of change and tensed existence, since tensed existence "entails that we can, at different times, rightly assert about the same thing that it exists and that it does not exist." [Here Gödel seems to be self-consciously turning Kant on his head, for Kant's famous remark is that "only in time can two contradictorily opposed predicates meet in one and the same object, namely *one after the other*" (1965, A32/B48).]

At first glance, this remark of Gödel's might seem merely to be based on an unfair criticism of a *tensed* concept of existence by employing, in the reductio clause, precisely a *tenseless* use of "exists." If our reasoning has indeed been along Gödel's lines, however, we can see that the explanation is rather that he seeks a notion of *reality* as *what exists,* in the sense of *all* that exists, and that it is the force of the "*all*" (which derives from the force of "reality") that drives his argument. The reply that tensed existence implies only that something exists *at one time* and does not exist *at another time* would surely not satisfy him, since he would then ask whether both *times,* in turn, really exist (are part of all that exists). If not, reality turns out to be an illusion (since it is incomplete). But if both do really exist, clearly *this* sense of "exist" must now be *tenseless,* and we will have spatialized time (in the sense characterized earlier), since position in time will now carry no ontological weight. And, as should be clear by now, for Gödel, to spatialize time *is* to render it ideal (by robbing it of its characteristic mode of existence).

If this reconstruction of Gödel's reasoning has been along the right lines, we can see why he cites the temporal idealist McTaggart as a precursor, since McTaggart's argument for the unreality of time has strong affinities with the version we have attributed to Gödel. The elusive details of McTaggart (1908) need not be gone into here, but the analogy with Gödel's approach is clear enough. What is crucial for our purposes is the tension between the *completeness* implicit in the classical concept of reality (Wittgenstein's "*all* that is the case") and the *essential incompleteness* that seemingly attends tensed, or temporal, existence. A reconciliation here seems difficult, indeed. There is always, as it were, something unfinished about the temporal. That is why it can never be complete, or completed, or completely described. Thus Michael Dummett can write, "McTaggart's argument shows that [if time is real] we must abandon our prejudice that there must be a complete description of reality" (1960, p. 357). And from an entirely different point of view, a relativistic one that yet takes the "dynamic" of time seriously, Čapek asks, "to what extent does the concept of the totality of the universe remain meaningful after the denial of absolute simultaneity of distant events?" (1961, p. 271). If we are not led by such considerations to give up, with Parmenides and Gödel, on the concept of time (or, more radically, the traditional concept of reality itself – Nietzsche? Heidegger?), then it would seem, as we hinted earlier, that there will at least be something ontologically suspect about the temporal – a kind of falling away from the fullness of genuine "being." And this seems to be the approach adopted by Plato, who tried, unlike his great predecessor, Parmenides, to save the appearances.

From three crucial dialogues, *The Sophist, The Parmenides,* and *The Timaeus,* we know that Plato was led by considerations like those we have discussed to distinguish (yet also to reconcile) the fundamental categories of *Being* (or Eternity) and *Becoming* – the former confined to the mathematical or formal realm (in Plato's sense of "form"), the latter to the phenomenal, to the world of appearances. And Plato, too, had a fundamental analogy for time, which he described, in *The Timaeus,* as "a moving image of eternity." As indicated earlier, "eternity," for Plato, stood not for a length of time, for "everlastingness," but rather for a *kind of being* – namely, *genuine.* "Eternity for Plato," as Harold Cherniss writes, "is not temporal duration; it is not a measure but a mode of being" (1962, p. 212). An "image," moreover, as Plato stressed in *The Sophist,* as a mere copy, can at most *resemble* and thus always fall short of its original. Temporal being is thus for Plato always a mere imitation of genuine being – what "is" in time is always "on the way to becoming" (what it really is). Thus, as J. M. E. Moravcsik puts it, "[g]iven Plato's views on the tenses, temporal being is not complete being; what is temporally both is and is not" (1973, p. 11).

Plato's conception of time had a strong influence on Aristotle's. Aristotle, in turn, distinguished actual from merely potential being (which is always on the way to becoming actual), and associated the latter with matter. The very essence, indeed, of matter is its potentiality, which it expresses by the "motion" of "becoming" (including, as motion, as Plato had, both locomotion and qualitative alteration). *Time, then, is what, in effect, separates the potential, qua potential, from its realization (always partial) as actual, expressed in the form of motion* (see Kosman 1969). To be fully actual, which is impossible for a material being, would thus also imply being eternal or outside time, and so not in motion. Arguing from premises such as that all potential being presupposes, for it to "awake" into its actualization, something fully actual, to awaken it, Aristotle concludes, in *Metaphysics XII,* that such a being does in fact exist. He calls it an "unmoved mover," or God.

These classical conceptions are, I would suggest, what Gödel presupposed as the background of (in his words) the "intuitive, pre-relativistic" idea of time against which he wished to place the limit case of relativistic space-time he had constructed. But the background will not be complete until we have sketched in an important idea of McTaggart's. McTaggart makes a considerable contribution to the analysis of classical accounts by introducing a fundamental distinction between two aspects of the temporal, which he calls the "A-series" and the "B-series." This distinction was implicit in classical philosophy, but McTaggart was the first to make it explicit and to give it a name. The B-series provides the fixed, immutable, indeed geometric relations between events in time in terms of *before* and *after.* The system of dates of events is B-theoretic, as are typical issues in the physics of time that are mathematically formulable, such as whether time is discrete or continuous, open or closed, and so on. The A-series, by contrast, characterizes events in terms of their shifting realizations as future, present, and past – that is, as *now.* Unlike the case with the B-series, an event's "location" in the A-series is not fixed once and for all but is always in flux due to the *nunc fluens* – the "flowing now." Thus, whereas it is a B-theoretic fact that I am writing this in 1991, it is an A-theoretic truth that it is now 1991, so that I am writing it now. The former will still be true fifty years from now; the latter, however, will not.

The A-series must be strictly distinguished from what one might call the "relative A-series." The relative A-series characterizes events as *now* only *relative* to a position in time in the B-series. Thus, in the relative A-series, whereas my writing this is now-for-me-in-1991, McTaggart's writing (1908) is now-for-him-in-1908. In a word, *every* time is *now* with respect to, or relative to, itself. With the relative A-series, then, no time is singled out as privileged – as *the* now. A moment's reflection reveals, however, that the relative A-series is in fact none other than the B-series! For "now-with-respect-to-time-*t*" signifies, surely, no more than "simultaneous-with-*t*." To keep McTaggart's distinction intact, then, it is important not to confuse the genuine A-series with the relative A-series, and it will prove crucial, in the pages to follow, that we keep these series separate.

Having distinguished the A-series from the B-series, McTaggart argued forcefully that for time to be real, both series had to exist. He also believed he had an argument that it is impossible for this to be the case, and so he concluded that time is an illusion. I alluded to this argument earlier in connection with Dummett's elaboration and defense of it (1960), but it is the distinction itself I wish to draw attention to, not the particular use to which it was put by McTaggart. Now, the B-series, as was suggested earlier, is amenable to a straightforward formal treatment, as in mathematical physics. But the A-series is more elusive, and a natural suggestion is that *it is this elusiveness of the A-series that is responsible both for the need we have pointed to, of Plato, Aristotle, Kant, and others to find an analogy for time, and for the ontological anomalies that seem to attend the temporal* – in a modest form for Plato and Aristotle and in an extreme way for the idealists Parmenides, McTaggart, and Gödel (and also, in a certain sense, for Kant). Since the question of the A-series would thus seem to be decisive for our investigations, let us attempt to characterize it more fully by indicating certain *concomitants* that would attend it, if it were realized.

(1) One time is distinguished or privileged – *the now* – and this privilege "passes on" to a succession of times, as time evolves or lapses. (2) Time evolves or lapses, irreversibly and in a certain direction – from the possible to the actual, from the future to the present to the past. (3) "Positions" in time are not "ontologically neutral"; the only location that confers

the privilege of existence is that of the present, the now. (4) Although the past is (now) closed and fully determinate, hence (now) "necessary," in the sense of "unavoidable", the future is still open. (5) Simultaneity is a nonrelative, two-term relation.

Now, if we consider just the Aristotelian conception discussed earlier, we can see why (1) to (5) would be seen as essential to the A-theoretic component of time. In fact, it is a rather delicate matter to separate out the A- from the B-theoretic components of time in Aristotle's account, but we need not dwell on such subtleties here. [See, on this point, the very helpful study by M. J. White, "Aristotle on 'Time' and 'A Time'" (1989).] We see, from this Aristotelian perspective, why the A-series has *ontological* significance, since it represents the coming into actuality (i.e., into actual existence) of the merely potential or possible. We see also why the A-series is successive and incomplete – since the material world, qua material, is always in the process of realizing itself – and also why it is "branchwise asymmetric" – since although the past is always determinate, and thus unique and linear, the future is indeterminate, "branching" into open possibilities. And as Aristotle also insists, the "flow" of successive times must not be taken literally, but, as we would say, "analogically," since *literal* motion or flow – say, of a river – occurs *in* time (and so at a determine *rate*). What, then, of the flow *of time?*

Clearly, the notion of *nunc fluens,* to be made sense of at all, must be regarded as primitive, as sui generis. Time may be, in some sense, *like* a river, but it is just where the analogy falters that intuitive insight must step in (if it can) to complete the picture. ["Hence Whitehead's conclusion that in analyzing time, an ultimate appeal must be made to intuition, that is, to our direct awareness of time" (Čapek 1961, p. 372)]. In a conceptual scheme, however, biased toward the visual and spatial, whose language (or logic, or *logos*) is *geometry,* how is one to recover what is not spatial but rather temporal? Not *sight* but *hearing* seems to be here the more appropriate sensory modality. Indeed, it is instructive to compare, on this point, the score of a musical composition, which is visually or "geometrically" graspable, with its successive realization in the very different medium of sound. Note further the different explanation needed for the unity of a melody, successive in time, from the nonsuccessive unity of a chord consisting of notes played, as it were, spatially – that is, together, nonsuccessively. Indeed, in their attempt to provide a set-theoretic foundation for music analysis, R. Dipert and R. Whelden (1976–7) define a chord as a (spacelike, as Gödel says) set of tones and a piece of music as an ordered *n*-tuple (sequence) of chords. [Later, we will discuss attempts to define ordered sets themselves in terms of (spacelike) unordered sets (an issue raised in Dipert 1982). It will emerge that such definitions sacrifice the temporality implicit in the intuitive conception of an ordered set or sequence.] If we accept, then, that a piece of music, either in itself or in its realization in performance as a succession of sounds, is in essence temporal, it remains that it can be represented in a spacelike musical score, and that one can, it seems, through a complete ('simultaneous') grasp of a score, succeed in grasping a temporal entity tenselessly. This puzzling situation is brought out nicely by Roger Penrose (1989, pp. 444–5). "Like Mozart, he [Bach] must somehow have been able to conceive the work in its entirety, with the intricate complication and artistry that fugal writing demands all conjured up together. Yet the temporal quality of music is one of its essential ingredients. How is it that music can remain music if it is not being performed in 'real time'? Penrose has, in effect, raised the question: Can one succeed in representing time or music purely conceptually (via the 'geometry' of a score), or must we, in the final analysis, construct (to use our earlier terminology) an 'analogy,' in the form, perhaps, of a temporal model? [Compare Brouwer: Can one represent genuine mathematical activity by the merely conceptual–logical 'geometry' of a (tenseless) formal system, or must one actually construct something mathematical in 'real time'? We will come to this question in the final chapter. For discussion of the characterization of a formal system like Frege's *Begriffschrift* as a kind of geometry, see Robert Tragesser (1984, p. 54): "if one simply looks at the pages of the 'Begriffschrift,' they have the character of presenting a rather beautiful geometry of signs. It would be absurd to say that in exercizing the logic presented in that work one was not relying on visual intuition."]

It is becoming clear, then, that there is a serious tension between the language of classical mathematics – with its logical space (in the general sense of this term discussed earlier) of numbers, lines, points, and so on – and the successive reality represented by the A-series. In the classical mathematization of time, points give rise to "instants," and in the relativistic mathematics, lines become world lines of particles; but *these* concepts are the natural means of capturing, rather, the *B-series*. This is, in fact, in part what underlies the problems with the reality of time brought out by Parmenides's student Zeno – for example, the paradox of the arrow in flight, which at no *instant* of its supposed motion is ever in an actual *state* of *motion* – which Plato discussed in his *Parmenides* and Aristotle in his *Physics*. [We will see later how Russell (1956) handles this problem.]

It would seem, then, that if one is to come to grips with the question of the reality or objectivity of A-theoretic time, one must, if a mathematization is sought, be open to a nonclassical systematization. The manifold that is generated by the temporal succession of the A-series cannot be represented by the pointwise decomposable continuum of classical analysis, but must be something more along the lines of the intuitionistic continuum of Brouwer. Moreover, as Čapek writes, "[i]t is unfortunate that the term 'continuity' in the mathematical sense in fact means infinite divisibility, *that is, discontinuity infinitely repeated* – the very opposite of continuity in the dynamical sense" (1961, p. 373; my emphasis).

The "now" of the A-series is thus *not* the problematic "knife edge" of a mathematical point. Indeed, as White (1989) shows, Aristotle himself got into trouble here, since his conception of the generation of times by a point seems to conflict with his own doctrine that time consists not of point-like instants but rather of "lengths," which are limited on *two* sides by "instants." That Aristotle's mathematical perspective was in general closer to the intuitionistic has often been noted. [See, for example, Lear (1980 and 1982).] Indeed, Aristotle, with Brouwer, (1) believed the natural numbers themselves to be generated in time (*Categories*, chap. VI); (2) accepted only potential, not actual, infinity (*Physics*, Bk. III); and (3) in his logic, cast doubt on the universal validity of the principle of the excluded middle (in Aristotle's case, the question of future contingent propositions caused him to reevaluate, if not to abandon, this principle) (*De Interpretaione*, chap. IX).

In sum, if the concomitants of the A-series are not merely to be dismissed out of hand, then one must resist the temptation to characterize them in a classical-geometrical sense. Čapek is correct to insist, then, in connection with consideration (2) and the so-called direction of time, that "[i]f we forget that the word 'direction' is borrowed from geometry and kinematics and therefore cannot be literally applied to time, we may thoughtlessly draw all the consequences of the deceptive analogy between 'movement of time' and 'movement in space' " (1961, p. 396). The direction of time – from potential to actual – cannot be captured in spatial-geometrical terms (which is why the question of the reversibility of time's direction is a nonstarter from the A-theoretic point of view – something cannot be "de-actualized" or "re-potentialized"). A too literal-minded conception of the motion of time would then, as Čapek says, present us with the question (meaningless for the A-theorist) of whether the future is approaching the present, or rather, whether the present is approaching the future, "as every motion in *space* is relative and can be transformed into a rest by an appropriate change of the frame of reference" (ibid., p. 346; my emphasis).

Collecting, then, our preparatory observations on the history of the concept of time, we remind ourselves that the paradoxes that Parmenides is said to have discerned in the attempt to reconcile genuine temporal becoming with a robust conception of reality gave rise to the ontologically fragile accounts of such becoming of Plato and Aristotle, from which McTaggart distilled the two dimensions of the A- and the B-series. Of these, it turned out, the A-series is the more formalistically elusive. Finally, not to render the A-theoretic conception immediately invisible, we insisted that the A-series not be conflated with the relative A-series, and that it also not be rendered meaningless by attempting an explication along the lines of classical analysis.

We can now state what lay behind Gödel's dramatic philosophical conclusions that he drew from his formal results on the GTR. Gödel believes firmly, I suggest, that (entirely aside from questions about the STR and the GTR, to invert Einstein's comment) if time is not to be a mere illusion, to be merely ideal, *it must include the A- as well as the B-series,* together with concomitants (1) to (5) of the A-series. Thus, Gödel (unpublished) states that "an objective lapse of time, such as is contained in the intuitive concept of time, is impossible in the R-worlds," as well as that "that passing of time which is directly experienced has no objective meaning in the R-worlds." Further, reinforcing the *ontological* and *nonrelative* character of A-theoretic time, he insists that, "[o]ne may say that . . . relativity theory has proved only the relativity of time not its subjectivity. . . . But it is to be noted that 1) the absoluteness (or uniqueness) forms itself an essential characteristic of intuitive time, and that 2) a relative passing of time (if any sense can at all be given to this phase) would be something radically different from the passing as described above; *for existence by its nature is something absolute*" (ibid.; my emphasis). Thus, the mere existence of a B-theoretic structure, satisfying whatever complex geometric relations you want, would be, for Gödel, insufficient to establish the reality of what is intuitively meant by "time": "[S]omething of this kind [which lacks the character of "passing by"] can hardly be called time" (ibid.; my interpolation).

If Gödel had meant by time, then, only B-theoretic time, his philosophical conclusions would be *incomprehensible.* In this respect, what Gödel demands of time in order for it not to be ideal stands in sharp contrast to the dominant perspective of contemporary philosophy of space and time. Among standard textbooks in the field, the A-series is either almost completely ignored, as in Bas van Fraassen (1985) and William Newton-Smith (1980), for example, or it is discussed only to be completely dismissed, as in D. H. Mellor (1981) and Paul Horwich (1987), none of whom, however, with Gödel, draw the conclusion from this that time is ideal. Indeed, in Mellor's *Real Time* the conclusion is explicitly drawn that B-theoretic time alone is *real* time! And John Earman writes that "time does not 'go,' or 'unroll,' or 'flow.' All attempts to give a literal-minded interpretation to the notion of the flow of time have met with failure" (1967, p. 217).

Needless to say, Earman does not deduce from this proposition that time is merely ideal, an appearance or illusion. But Gödel does. "One may take the standpoint," Gödel writes, "that the idea of an objective lapse of time (whose essence is that only the present really exists) is meaningless. But this is no way out of the dilemma; for by this very opinion one would take the idealist viewpoint as to the idea of change, exactly as those philosophers who consider it as self-contradictory. For in both views one denies that an objective lapse of time is a possible state of affairs, a fortiori that it exists in reality" (1949, fn. 4). (I take it that Gödel's "lapse of time," as well as "passing by," from our earlier quote, are equivalent to Earman's "go," "unroll," and "flow.") Note, further, that Earman is unhappy with the lack of a "literal-minded" account of this A-theoretic notion. As we have seen, however, all A-theorists would in fact insist that an analogy is needed here. What is unclear is why Earman seems to reject out of hand any employment of analogical reasoning in this area. Is it that, for him, *whatever* is real *must* admit of only a literal-minded interpretation? Whatever the answer, it is sufficiently clear that Gödel imposes stringent requirements on time for it not to be ideal, exactly in the tradition of Parmenides, Kant, and McTaggart, whom he correctly cites as his philosophical precursors. What led him, then, to conclude from his formal results on relativity theory that it is doubtful that these stringent requirements on time can be met?

Gödel (unpublished) speaks of an increasing "objectivation" of the temporal in the history of the natural philosophy of time. More and more features attributed to time in the intuitive conception have been eliminated from the scientific world-view as having, at most, "subjective" significance. Thus, as it were, the class of those intuitive features of time that have physical significance has been steadily shrinking as physics has progressed. Indeed, time is not unique in this regard. The road to the contemporary mathematization of nature was paved, after all, by Galileo's famous introduction of the distinction between primary and secondary qualities, which, by relegating the latter to the status of the merely subjective, freed the natural philosopher from the need to take them into account in his mathematical formalizations. Thus, Lawrence Sklar, in "Up and Down, Left and Right, Past and Future" (1985, chap. 12), echoing, in effect, Gödel, speaks of modern science proceeding via a "secondarization" of the phenomena. How else, for example, could one be asked to accept identities such as Heat = Mean Kinetic Energy unless one distinguished heat, in this context, from the sensation of heat – that is, a mere secondary quality? Sklar questions whether this strategy will work in recent attempts to capture time in purely causal or entropic terms. The disanalogy occurs, for him, because although heat (as a physical process) is in essence independent of the sensation of heat, the experience of temporal succession is not similarly independent of the existence of genuine succession in time (at a minimum, even a succession of subjective experiences is a genuine, and hence objective, succession, and so implies the existence of time). And Saul Kripke (1980), it will be recalled, argued along similar lines, stating that a mind–body identity theorist who asserts that pain = *c*-fiber stimulation can hardly, in the spirit of Galileo, ignore the merely secondary qualities of pain, since the sensation of pain constitutes the very essence of pain itself.

Gödel (unpublished) points out that by the time of Kant's reconstruction of the then contemporary scientific worldview, "these steps or levels of objectivation, each of which is obtained from the preceding one by the elimination of certain subjective elements," had proceeded to an advanced stage. "The "natural" world picture, i.e., Kant's world of appearances itself," Gödel writes, "also must of course be considered as one such level, in which a great many subjective elements of the 'world of sensations' have already been eliminated." Kant, of course, was at pains to distance himself from the empirical phenomenalist Berkeley, who refused to relegate the secondary qualities of nature as experienced to a merely ideal status. Yet, for all that, as Margaret Wilson has argued, "Kant can be styled a phenomenalist not because he accords 'reality' to the ordinary 'image,' but only because the scientist's sophisticated reductive explanation of that image is, on Kant's view, elaborated within the framework of merely ideal 'forms of intuition'" (1984, p. 170). This peculiarly Kantian "transcendental" idealism, however, is too much for Gödel, who sees in it an overgeneralization of what is an otherwise useful distinction between secondary and primary qualities, or the subjective and the objective. "Unfortunately," he writes, "whenever this fruitful viewpoint of a distinction between subjective and objective elements in our knowledge (which is so impressively suggested by Kant's comparison with the Copernican system) appears in the history of science, there is at once a tendency to exaggerate it into a boundless subjectivism, whereby its effect is annulled. Kant's thesis of the unknowability of the things in themselves is one example" (ibid.).

In fact, Kant's position on the question that is of particular interest here, the experience of time, is rather slippery. On the one hand, he insisted that the reason we have (or so he believed) an a priori insight into the fundamental features of time is that (as we saw earlier) "[t]ime is nothing but the form of inner sense." On the other hand, he insisted, in the Second Analogy, that our subjective time-consciousness is somehow grounded in our ability to order events in the world in a lawlike way in terms of cause and effect. "[T]he subjective succession by itself is altogether arbitrary" (1965, A13). But is an arbitrary succession a succession? The delicacy of this question arises from the attempt to discern genuine, that is, objective, properties of the subjective realm, qua subjective, and the difficulty one has answering it brings to the surface the fragile nature of Kant's distinction between subjective and objective elements in experience. (We will come back to this question later.) It is no accident, then, that when these two strands in Kant's philosophy, already in tension, are pulled apart, as they are by recent advocates of the causal theory of time, no doubt inspired by the "causal time" (in the sense we alluded to earlier) of Einstein's relativity theory, trouble ensues. [For some insight into these troubles, see Newton-Smith (1980, p. 204), Whitrow (1980, pp. 225, 323, 326), and Sklar (1985, pp. 256–8).]

The Kantian level of objectivation is, for Gödel, exceeded by the STR, where, he suggests, time has become dramatically more spacelike. Absolute global simultaneity, in the non-frame-relative, intuitive, two-term sense, has disappeared. Gödel interprets this theory philosophically in the same way Einstein does, which we indicated earlier, as "replacing the *evolution* of a three-dimensional existence" with a "rigid, four-dimensional *space.*" Some, however, as we will see later, including Čapek and Howard Stein, prefer a dynamic reading of the STR, in contradistinction to Einstein and Gödel. Whitrow (1980) is more circumspect. His view is that "the passage of time corresponds to the continual advance of the light cones" (p. 348), and he asks us to "visualize . . . a complex of advancing light cones in space–time, the track of each vertex being the world-line of a potential observer" (p. 352). Nevertheless, he is careful to observe that "[a]s far as the theory of relativity is concerned, we can consider *either* the *set* of *all* these light cones *or* the *continual transition* from one to another" (p. 348; my emphasis). Gödel, however, as we have seen, attributes to the "passage of time" a strong ontological force that imposes on the *now* a strict requirement of uniqueness, and thereby conflicts with the relativity of simultaneity (and therefore of the *now*) in the STR. Therefore, he is already suspicious of attempts to introduce the passage of time into the STR, and his new model of the R-universe, by his lights, deals the coup de grâce, from the perspective of the GTR, to any possibility of reconciliation between relativity theory and temporal becoming.

In relativity theory, then, as Gödel reads it, there is no objective correlate of the subjective experience of the passing of time. There is, to be sure, a spacelike "trace" of the subjective, or intuitive, experience of temporal becoming in the definite, objective ordering of events "from past to future" determined by the light-cone structure of Einstein–Minkowski space–time. But there is, as Gödel sees it, no objective *temporal becoming* correlate with (the) subjective (experience of) becoming. Time appears, rather, as a kind of fourth ('spatial') dimension, *distinguished,* to be sure, from the other three, but *only geometrically,* in terms of the world lines permitted by the light-cone structure of Minkowski space–time and so *not intuitively,* as Kant had distinguished time as a form of *intuition* distinct from space. Those, therefore, who base their opposition to the view that the STR spatializes time on the fact that, in the equation for the interval, the time variable is distinguished from the spatial by being preceded by a negative sign have not really addressed Gödel's concerns. Newton-Smith (1980), for example, unimpressed by the "range of *alleged puzzles* that some philosophers have thought to arise because of *supposed tension* between the STR and our ordinary conceptions of past, present and future" (p. 187; my emphasis), insists that "[t]he theory maintains an absolute distinction between space and time. This is reflected by the fact that the time coordinate in the semi-Euclidean Minkowskian metric has the opposite sign to the spatial coordinates: $ds^2 = dx^2 + dy^2 + dz^2 - dt^2$" (pp. 186–7).

Gödel's conclusions, however, are unaffected by these facts about the *geometric constraints imposed on Minkowski space–time by this definition* – by the fact, as it were, that time here has a different *geometry* from space. His reading of Relativity Theory speaks not to these B-theoretic facts, but rather to the A-series of temporal becoming. Indeed, Newton-Smith seems insensitive to just this distinction – crucial, I have suggested, to Gödel's thinking – since, as was seen earlier, after citing the above B-theoretic constraints, he goes on to cast doubt on the "supposed tension between STR and *our ordinary conceptions of past, present and future.*" In fact, far from acknowledging the negative sign before the time variable in the above equation as the signature for the "absolute distinction between space and time," Gödel views it rather as "the last objective remnant of time," which he well expects to disappear as physics reaches yet another level of objectivation. "The one-dimensionality of time," he writes, "it is true, expresses itself also . . . in the fact that the fundamental quadratic form has one negative square, but . . . it would only be a straightforward continuation of the development from classical physics to relativity theory, if this last objective remnant of time too were to disappear" (Gödel unpublished). And Gödel's view of the matter, as opposed to Newton-Smith's, is echoed in the recent book by Penrose (1989, pp. 443–4):

"The way in which time is treated in modern physics is not essentially different from the way in which *space* is treated and the time of physical descriptions does not really 'flow' at all; we just have a static-looking fixed space–time in which the events of our universe are laid out! . . . It is hard to believe that what makes time 'actually flow' in our experiences of the physical world we know is merely the asymmetry between the number of space dimensions (3) and time dimensions (1) that our space–time happens to have."

In spite of the high level of objectivation that Gödel discerns in the STR, however, there remains that "last objective remnant of time," as well as the fact that from the perspective of the GTR, given an appropriate distribution of matter and motion in the universe, one can still define a "cosmic time." The objectivation of time, then, is not complete. To use our earlier terminology, in the theory of relativity we do not yet have a limit case in the mathematization of time. The result of this, for Gödel, is that the relationship of the formal to the intuitive elements of time is not yet fully visible, which prevents us from being able to recognize the striking and counterintuitive philosophical consequences implicit in Einstein's theories. What Gödel achieved, however, in his rotating universe solutions to the gravitational equations of the GTR is precisely a limit case for the relativistic mathematization of time. For in the R-universe it is provable that no cosmic time is definable, and it becomes impossible, by Gödel's lights, to continue to maintain the existence of any semblance of an objective lapse of that (genuine, successive) time that one might have held out for in the STR. The latter derives from the fact that in the R-universe, by traveling always toward one's causal future, one can arrive eventually at one's local past, thus "canceling," as it were, the effect of the passage, the objective lapse, of time.

In such a universe, one encounters *circular* or *closed* time, in contradistinction to *cyclical* time, in which a temporal process is *repeated* (*n* times). Even though the past is over, then, one somehow returns to it, thus making nonsense of the idea of a genuine *succession* of times, in the temporal (A-theoretic) sense, in which only the present member of a succession exists. In a temporal circle, as opposed to a temporal cycle, one does not *repeat* a series of processes, as a second hand completes the circuit of a dial so many times. For a temporal cycle, one can thus always ask how often the circuit has been completed, whereas such a question has no meaning with respect to the spacelike existence of a time circle. (Recall our earlier definition of a space in the general sense. If one can travel, at will, to any part of time, this must be because all parts of time, like all parts of physical space, *already exist* to be traveled to.) The crucial question, then, is whether one can sustain the philosophical consequences that Gödel derives from his new model for the GTR.

3

Time travel and the Gödel universe

> Once Einstein said that the experience of the Now means something special for man, something essentially different from the past and the future, but that this important difference does not and cannot occur within physics.
>
> Rudolph Carnap, *Autobiography* (Schilpp, ed. 1963, pp. 37–8.)

In order to grasp the significance of the presence of nearly closed and closed timelike curves in the Gödel universe, it is useful to consider the following two representations. Figure 1 illustrates the successive stages in the advance of "intuitive" time, and Figure 2 shows what would have to obtain if closed time were possible. (A) in Figure 1 can be easily seen to represent a succession of time. In the first place, it unfolds stepwise, successively. Further, when an earlier step *has* occurred, the later steps *have not* (they lie in the future, hence they are not yet actual or existent). Finally, it is always appropriate to ask, at any given step in the temporal progression, what stage *has now* been reached – and there will always be a definite, unique, answer.

None of these conditions, however, makes any sense when applied to (B), in Figure 2, considered as a representation of closed time. For in such a model of time, stage (5) no more comes about only after stages (1) to (4) have – that is, waits upon their completion for its own existence to be realized – than they wait upon its completion. Rather, it is entirely irrelevant, from an ontological point of view, how one chooses to describe the order of completion of the different steps, simply because they are all equally complete at all

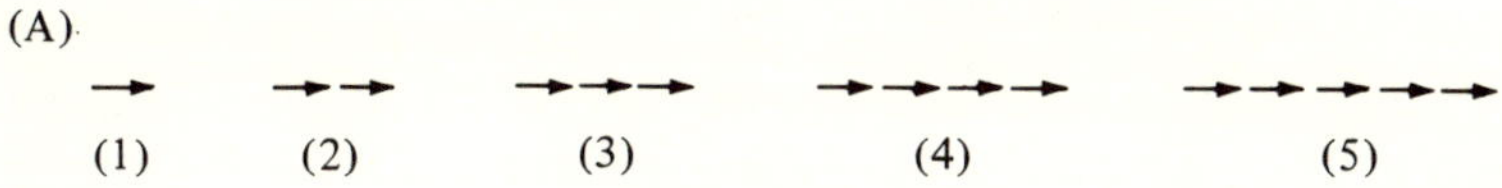

Figure 1

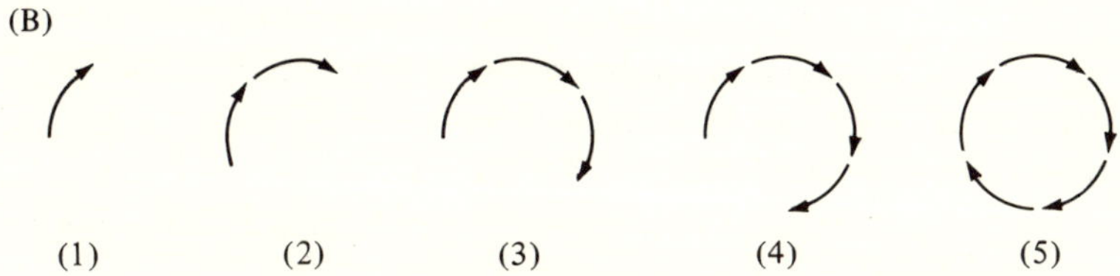

Figure 2

stages. If the step at stage (5) were not already fully complete at stage (1), for example, step (1) could never occur, since it occurs after step (5). This is what it means to describe time as forming a circle. In a world with closed, circular time, then, time could not but be spacelike – that is, temporal position in it could not have the ontological significance that attends a genuine, open succession of stepwise generated times. *And that was, of course, exactly Gödel's point.* That is why he believed that the new cosmological model he had discovered confirmed the idealistic conception that there is no "objective lapse of time." "[T]he decisive point is this," he writes, "that for *every* possible definition of a world time one could travel into regions of the universe *which are passed* according to that definition. This again shows that *to assume an objective lapse of time* would *lose every justification* in these worlds" (Gödel 1949, p. 264; first emphasis, Gödel's).

Further confirmation of Gödel's thesis comes from the fact that all five concomitants of the A-series, as we characterized them earlier, disappear in the closed time of the Gödel universe. Only a conflation of the A- with the B-series, then, could lead one to say, with Richard Sorabji, in "Closed Space and Closed Time" (1986), that, "[d]espite the fact that each event could be seen as lying equally in the future and in the past, I see no reason why there should not be a *direction* for the *flow of time*" (p. 221; my emphasis). Consider, however, the concomitants of the A-series. (1) There would obviously be, in closed time, no unique, privileged, *now.* (2) Time would not "flow" or "lapse," for the reasons given earlier. Further, although one could specify a geometric, B-theoretic 'direction' for time in such a world, this would not be the direction that characterizes A-theoretic time, which moves from the possible to the actual. For time travel to the past would be travel to what was already actualized and cannot, therefore, be re-actualized. Indeed, the very distinction between actual and possible times breaks down in the case of closed time. (3) Positions in circular time would be, as we saw, ontologically neutral. (4) The future would be as determinate and complete as the past, and therefore not open but closed. (And, as we will see later, in connection with Aristotle and the logic of indeterminist time, this would mean that if the past is 'necessary', then so, too, is the future.) Finally, concomitant (5) clearly fails, since the closed time of the Gödel universe is part of a relativistic model that presupposes the relativity of simultaneity.

To drive home, further, the force of his conclusions from the peculiar properties of the R-universe, Gödel refers to certain paradoxes that he says would follow if it were feasible in such a universe for one "to travel into the near past of those places where he has himself lived" (1949, p. 262). He proceeds to argue that in fact such travel would not be practically feasible, and so "[t]herefore it cannot be excluded a priori, on the ground of the argument given, that the space–time structure of the real world is of the type described" (1949, p. 264). We will come in a moment to the question of whether these 'paradoxes' of time travel to one's local past amount to actual antinomies. Gödel's point, here, however, is that factors such as these should not lead us to rule out the R-universe as not containing physically realistic space–time structures, structures in which, nevertheless, no sense can be made of the objective passage of time. It is well, then, to remind ourselves of the two-faced nature characteristic of Gödel's formal results, which we drew attention to earlier.

Gödel cites certain mathematical results he has derived in relativistic cosmology concerning the possibility of the R-universe, results that *would* lead to the conclusion that *time* travel, in an intuitive sense, is possible in such worlds *if* we assume that *t* still has, in these contexts, the *standard* interpretation as *time*. Precisely this standard interpretation, however, Gödel rejects, arguing, as we have seen, that nothing corresponding to the passage of time (which he sees as essential to the existence of time and change) remains in such a universe. Thus, what could otherwise be correctly described as *time* travel – namely, the existence of nearly closed and closed, timelike world lines – cannot, in strictness, be said to satisfy this description, since exactly those conditions that *permit* us to describe these world lines as nearly closed or closed *prevent* us from also characterizing them as *temporal*. In a word, just because there is, loosely speaking, 'time travel' in such worlds, there is no passage of time, hence no time in the intuitive sense, and therefore no genuine time *travel*. (Indeed, there is no travel of any kind, in the intuitive sense, in such worlds.) Certain unexpected formal structures of space–time would exist, to be sure, but a certain interpretation of them – the standard one – is, for Gödel, definitely ruled out.

What remains to ask of such a model are B-theoretic questions, as well as issues concerning how physically realistic we should take this hypothetical universe to be. We will come to the latter issue presently, but let us focus for now on the 'paradoxes' of time travel that may seem to arise in the context of B-theoretic time. The first issue to be clear on is that a time traveler will not be remaking history. The very reason he or she will be able to visit the past is that in the spacelike B-theoretic time of a closed-time structure like the Gödel universe, all parts of time are equally real, existent, or actual. If it is correct, therefore, to say that one will visit the past, then it is equally correct to say that one already has visited it. As we saw earlier, a temporal circle is not a repeated cycle, and so one is not *re*visiting the past. One cannot 'undo,' therefore, anything in the past; if it has already happened, it is a fact, and nothing – including time travel – can alter the facts. Indeed, the most prominent B-theoretic consequence of time travel in the Gödel universe is *the symmetry of past, present, and future with respect to facticity.* The asymmetric time structure of an A-series "open" future, pregnant with multiple possibilities, is not applicable here, in a world containing only actualities, and so nothing can be *changed* – neither events in the future nor those in the past (even if one travels to it). Thus there is no danger that the time traveler might, by accident or design, kill his or her parents and thereby eliminate the conditions necessary for his or her own existence. The very existence, rather, of the time traveler guarantees that no such result will come about – either because, as Gödel suggests, such a trip would turn out not to be feasible or because the time traveler on such a trip to his or her local past somehow, in fact, always fails to produce results that, if successful, would prevent his being there.

This last hypothesis brings out the fact that closed, circular time imposes strict constraints on what events can occur (past, present, or future), consistent with the fact that certain other events have happened. The 'beginning' presupposes the 'end,' as it were (for whatever point in time we arbitrarily choose to call the beginning). What is possible to do now, thus, depends on what was actual before. But, as we observed earlier, there are really no *genuine possibilities* in such a universe, and hence no free will in the traditional sense in which this implies the possibility of doing otherwise. In a universe of pure fact, containing only actualities, the most severe requirement is one of logical consistency – a requirement that reality, of course, unlike a formal system, cannot fail to satisfy. It remains to determine how best to characterize requirements such as that, if I try to kill my "earlier self," I will fail. Either there would always be some ad hoc reason why such an action would fail (for example, the gun would jam, a bolt of lightning would startle me, etc.), or (Horwich 1987) there would be in such a universe " . . . the sort of initial conditions that will give rise to certain types of coincidence" (p. 126). Indeed, in the Gödel universe, with the consistency conditions mentioned earlier, the distinction between mere initial conditions and laws of nature may break down, for, as Sklar has observed, "[i]f there is, relative to the propositions we now countenance as the laws of nature (including the laws of general relativity) a *lawlike* constraint on 'initial conditions,' if, that is, some specifications of the world 'at a time' are impossible since they are lawlike incompossible with themselves, what solid distinction is left between laws of nature and 'mere matter of fact initial conditions'?" (1984, p. 106).

In addition to these metaphysical-logical issues concerning the peculiar B-theoretic properties of time travel, however, there are certain epistemological anomalies to which Gödel has drawn attention. What troubles him is that if time travel to one's own local past were possible, one could encounter one's own past self, and "[n]ow he could do something to this person which by his memory, he knows has not happened to him" (Gödel 1949, p. 264). (Notice that Gödel does not say: "he thinks he knows.") The problem here is not simply that our memories might prove to be fallible. This situation arises already in the actual world. Rather, Gödel is worried by the fact that our decision-intention to travel through time and create mischief can be the *basis* for the disconfirmation of even a confident first-person memory claim. My memory, it would seem, could then, in principle, if time travel to such regions were possible, be revealed as spurious because of what I now *decide* to do.

This is indeed awkward and unexpected, but Gödel does not specify if this is possible for any category of memory report (so that no class is privileged), or whether it is possible that by this means we could, in principle, choose to undermine most of our memory claims. Though such possibilities would not issue in explicit antinomies, they would raise serious questions about the epistemological status of memory in such a world. (And the relationship between time consciousness and memory, of course, is intimate. See, for example, Newton-Smith 1980, pp. 205–96.) The anomalous character of the situation Gödel envisages can be brought out more fully if we recall that [as Dummett has emphasized in "Bringing about the Past" (1978a)] normally we take it to be a mark of the difference between past and future, between memory and anticipation of the future, that only for the latter is it necessary to take into account, in forming our judgments, our intentions. That is to say, our knowledge of what will come about in the future, as opposed to what occurred in the past, may be influenced by what we presently intend to bring about. Of course, the reply may be made that such anomalies are to be expected here, since, as we have seen, in a Gödel universe permitting time travel, our traditional conceptions of past-future asymmetry would be inapplicable. Perhaps, then, there is a standoff here between a view like Gödel's, which points out the strain that certain trips to one's own immediate past would put on epistemological categories like memory, and the contrary conception that, in a time-anomalous world like the Gödel universe, it is to be expected that traditional epistemic notions like memory will not survive unchanged.

It remains to ask, however, given the close relation, mentioned earlier, between memory and time consciousness, whether what remains of time, in a world permitting Gödel's 'paradoxes' of memory during time travel to one's own local past, is close enough to our own conception to permit the continued use of the term "time." One is tempted to remark, with Earman, that "one must face up to the question as to what extent the concepts of time, causality, and identity which appear in a coherent account of time travel would resemble our present concepts" (1967, p. 221). As we have seen, however, it is precisely Gödel's point that time in the intuitive sense would not exist in such a world – not for the reasons Earman adduces, but rather due to the fact that (for Gödel) no sense can be made, in such a universe, of the objective lapse or passage of time (a consideration whose force must surely be lost on those like Earman, who, as we saw earlier, can in any case make no sense of the notion of the flow of time even for our universe). Thus, as should be clear by now, there is a sense in which Gödel was not suggesting the possibility of genuine *time* travel in the Gödel universe. His dialectical use of his nonstandard model of the GTR was intended, rather, to deliver the coup de grâce to the notion that relativity theory is in the end compatible with the intuitive conception of the temporal – according to which there exists an objective passing of time. "That at least *that* passing of time which is directly experienced has no objective meaning in the R-worlds also *follows from the fact* that it is possible in these worlds to travel arbitrarily far into the past and future and back again" (Gödel unpublished; second emphasis, mine). If he continues, nevertheless, to employ the *term* "time travel," this is surely because he is confident that the reader will not forget his over-arching point that all temporal phenomena in the Gödel-universe are *ideal*.

In deducing the possibility of a universe permitting time travel, then, Gödel was, in effect, generalizing the geometrization of the temporal even to conclusions that Einstein himself was reluctant to draw. Einstein had always denied the possibility of "sending wireless signals into the past," and in relativity theory, as it is usually understood, this is indeed impossible. The reason for this impossibility, however, concerns the structure of Minkowski space–time, not, as in Kant, because of the category distinction between (the "forms of intuition" of) space and time. This extreme geometrical approach to time left the door open to Gödel's unexpected proof that certain counterintuitive geometrical properties of time will obtain in universes consistent with the purely formal–mathematical constraints of the GTR. Thus Einstein can write, in response to Gödel, that "[t]he problem here involved disturbed me already at the time of the building up of the general theory of relativity, without my having succeeded in clarifying it. . . . Does it make any sense to provide the world-line with an arrow, and to assert B is *before* P, A after P? Is *what remains of temporal connection* between world-points in the theory of relativity an asymmetrical relation. . . ?" (Einstein 1949, p. 687; second emphasis, mine).

Couldn't Einstein have replied, however, that one might be able to rule out the R-universe as realistic on the basis of purely physical considerations? He could have, and he did. "It will be interesting to weigh," he writes, "whether these [the Gödel solutions] are not to be excluded on physical grounds" (ibid., p. 688). This reply, however, does not address the modal force of Gödel's philosophical argument (alluded to at the opening of Chapter 2). Gödel seems to be arguing that the mere mathematical *possibility* of an R-universe, in which time *would* be ideal, shows that in our actual universe time *is* ideal. Indeed, as was mentioned in the Preface, I have suggested, in my "Review Essay: Hao Wang, *Reflections on Kurt Gödel*" (1989), that this pattern of modal reasoning – namely, that possible existence has strong implications for actuality – is typical of Gödel. In Wang's *Reflections on Kurt Gödel* (1987), for example, we read that Gödel had formulated his own version of the ontological argument for God's existence (whose characteristic logical form is that if God is possible, then He is actual). And for a mathematical Platonist like Gödel, the mere possibility of a formal structure would, I take it, imply its actual (mathematical) existence.

In the present instance, however, this characteristic pattern of modal reasoning is more elusive. To begin with, in order to have implications for the actual physical universe, Gödel's hypothetical cosmological model must be taken to have realistic, physical significance. On just this point, however, cosmologists have been skeptical. Georg Kreisel, for example, reports, concerning one prominent cosmologist, that, "by Professor Penrose's account, the direct physical interest of Gödel's papers . . . is limited – in accordance with Einstein's comment on [Gödel (1949)] . . . Professor Penrose emphasizes from his own work experience . . . [that] Gödel's work served as a cross check on mathematical conjectures and proofs in the modern global theory of relativity" (Kreisel 1980, p. 215).

It is tempting, then, to respond to the question at hand as Stein has: "Gödel's results concern a cosmological model; in assessing the eligibility of that model, one is up against the well-known difficulty that only one cosmos is given us (so that what other ones are 'physically possible' is moot)" (1970, p. 594). Although it can hardly be denied, however, that the question of physical possibility is "moot," it is not clear to me what force lies behind the suggestion that "only one cosmos is given us." It is certain that only our actual cosmos is given us via our sense experiences. But can we not grasp, conceptually, merely possible universes? In fact, does not the very enterprise of the theoretical scientist involve confronting the actual world with an ensemble of models of possible universes (often highly imaginative models – such as Einstein's), which it is the duty of the experimentalist to confirm or refute? Does not the experimentalist, then, by comparing, as it were, such models with the cosmos that is given us by sense experience, try to determine which universe is physically *actual,* as opposed to which are physically *possible* – the latter task falling, rather, to the theoretician? Although I have nothing new to offer the theorist concerning this general characterization of his or her task, it seems to me that something can be said concerning the specific question now before us – namely, the physical significance of Gödel's hypothetical world model.

The first thing to note about the Gödel universe is that it appears, to Gödel, at least (1949, p. 265), not to violate the natural laws that obtain in the actual world. (Stein 1970, pp. 593–5, discusses this issue.) One might, then, suggest that we should recognize a new law of nature – an anti-Gödel law, as it were – that would rule out the embarrassing situation in which the R-universe might turn out to be lawlike compossible with ours. But the question is: What might motivate one to entertain such a new law – except for the fact that the Gödel universe appears to be counterintuitive in implying the existence of time travel? It is, however, as we have seen, precisely one of Gödel's points that the full counterintuitive implications of relativity theory have not always been sufficiently realized – implications that his new models of the GTR were designed to bring out. Indeed, as we have seen, Gödel viewed relativity theory itself as but one (advanced) "level of objectivation" of time, to be followed, he suspected, by other, higher levels – where, he conjectured, even the "last objective remnant of time" will eventually be "generalized away." Although those who are skeptical of the physical significance of Gödel's results, then (viewing them, perhaps, as Penrose suggests, merely as "a cross check on mathematical conjectures and proofs"), are arguing by *modus tollens* (against the physical significance of Gödel's extrapolations from Einstein), Gödel surely sees himself as proceeding by *modus ponens* (bringing to light, as we said earlier, philosophical consequences not nearly so visible under less extreme geometrical constructions).

To decide the issue between *modus tollens* and *modus ponens* here, it is necessary to refer back to the various considerations we have brought attention to in this book concerning Gödel's philosophical interpretation of relativity theory. Earlier, with regard to his work on the foundations of mathematics, we pointed out that he had observed, in a letter to Wang about his incompleteness results, that a contrary philosophical approach had hampered the formalists (who "considered formal demonstrability to be an *analysis* of the concept of mathematical truth and, therefore, were of course not in a position to *distinguish* the two"). Here, I would suggest, Gödel presupposes for his formal results not to be robbed of their full significance, the philosophical account of relativity theory he has put forward. Concerning this account, we have already tried to reconstruct some of Gödel's reasoning. Of particular note is the fact that the problematic characteristics of time travel in the Gödel universe that concern the canceling of the passage of time by visiting the past, and belying the unreality of the future by visiting it at will, as well as casting doubt on the notion of the asymmetric direction of time, in the intuitive sense, all relate to *A-theoretic* time. And if Gödel is right, philosophically, *this* has, irrespective of considerations to do with time travel, already disappeared from the point of view of the *STR*. We will attempt to reinforce this reasoning soon, when we come to discuss those who, contra Gödel, have adopted a less counterintuitive interpretation of relativity theory. Assuming for now that Gödel's interpretation has force, and thus that his new mathematical model for the GTR has physical significance, it remains to spell out the reasoning by which he drew conclusions about the actual world from these results about merely hypothetical, possible worlds.

Gödel's modal reasoning, then, seems to be as follows. Since the actual world is lawlike compossible with the Gödel universe, it follows that our direct experience of time is *compatible* with its ideality (assuming, with Gödel, its ideality in the Gödel universe). But if even *direct experience* is inadequate to establish the existence of intuitive time – that is, not merely (relativistic) causal or cosmic time, but genuine, successive time that lapses or passes – then *nothing further* will suffice. Thus there are, and can be, no grounds for asserting the existence of intuitive time, of temporal becoming. In a word, "if the experience of the lapse of time can exist without an objective lapse of time, no reason can be given why an objective lapse of time should be assumed at all" (Gödel 1949, pp. 264–5). This line of reasoning shows a striking similarity to Kant's arguments in the Transcendental Deduction and Refutation of Idealism. In very simplified form, these arguments assert that if our possession of the basic categories of experience is compatible with the possibility that all our experience (as) of the external, sensible world of appearances is an illusion, then nothing further can establish that we are not subject to such an illusion. Thus, in effect, if there is no external, sensible reality in our world as we experience it, then there is none in any other world. (So, for Gödel, if there is no objective lapse of time in our world as we experience it, then there is none in any other world.) Kant claims, along these lines, to refute empirical idealism and to 'deduce' transcendental idealism.

Now, there would appear to be, for both thinkers, a movement from skepticism (both p and not-p are compatible with all possible experiences) to idealism, and as Strawson (1985, p. 16) notes, this kind of move rests, ultimately, on some sort of 'verificationist' bridge principle (Strawson speaks of Kant's 'principle of significance'). In the case of the STR, Gödel (unpublished) makes such a principle explicit: " . . . in perfect conformity with Kant, the observational results by themselves [on which the STR rests] really do not force us to abandon Newtonian time and space as objective realities, but only the observational results together with certain general principles, e.g., the principle that two states of affairs which cannot be distinguished by observations are also objectively equal" [my interpolation]. [Compare with J. S. Bell ("How to teach Special Relativity", 1989, p. 77): "Since it is experimentally impossible to say which of two uniformly moving systems is *really* at rest, Einstein declares the notions 'really resting' and 'really moving' as meaningless. . . . Lorentz, on the other hand, preferred the view that there is indeed a state of *real* rest, defined by the 'aether,' even though the laws of physics conspire to prevent us [from] identifying it experimentally. The facts of physics do not oblige us to accept the one philosophy rather than the other".]

For Gödel, then, to return to our earlier point, what is timelike (in the intuitive, not the relativistic, sense) about relativity theory has, at best, only an accidental correlation in the actual world with the intuitive features that make time *time.* By projecting pure 'Einsteinian time,' however, into a universe where it survives, but where the accidental correlations with intuitive time disappear, Gödel believed he had shown that even in our world (where the obtaining of these correlations renders less visible the true temporal profile of relativity theory) Einsteinian time is only a formal simulacrum of intuitive, successive time.

With this line of reasoning before us, then, we can begin to see why Gödel (unpublished) is primarily a long philosophical commentary on (in his words) the "insufficiently appreciated" similarities between relativity theory and Kantian (idealist) philosophy. (He is careful to add, however, that "I wish to say right in the beginning that I am not an adherent of Kantian philosophy in general.") "Kant and Relativity Theory," Gödel writes, "go in the same direction, but Relativity Theory goes only one step" (ibid.). Thus, just as in the STR 'causal' time provides simultaneity relations only relative to each inertial frame – or, if you wish, to each observer – so, in Kant's philosophy, time as a mere form of our sensibility represents only a relation between us and things-in-themselves. In both, time is ideal in that it is not intrinsic to the facts, but rather is imposed on them by an observer. "[I]t [namely, time] does not inhere in the objects, but merely in the subject which intuits them" (Kant 1965, A38; my interpolation). (Gödel remarks, however, on this difference, that for Kant time is relative to the universal human sensibility as such, not, as is the case with relativity theory, to particular inertial systems, or observers.)

Relativity theory "goes only one step," however, because

(1) the time relative to the co-moving frames can, compatible with the GTR, be fitted into a single, consistent, 'cosmic' time (which then is not merely imposed by each observer in virtue of his local frame of reference), and (2) even in the STR, the relation of temporal succession between timelike separated events is invariant (and so, again, is not merely imposed by an observer). In Kantian philosophy, however, there is no higher perspective from which time can be seen not to be ideal. Beings with a different sensibility from ours, Kant affirms, would not experience the world temporally at all. Here Kant's idealism is closer to Gödel's, since for the Gödel universe (and hence, so the argument goes, for ours) nothing satisfying the intuitive (Kantian, as Gödel puts it) characteristics of time can be realized. Yet Gödel's idealism, of course, unlike Kant's, rests (in the present context) on an empirical physical theory – the relativity theory of Einstein – and, again unlike Kant's, applies only to time, not to space (or space–time). Can Gödel's employment of relativity theory, however, for these idealistic purposes be defended against those who find themselves able to reconcile the theory of relativity with temporal becoming? We have attempted to trace the lines of Gödel's modal argument from the GTR and cosmology. It remains to evaluate Gödel's reading of the STR.

4

Not everything can be relativized

> The concept of existence, however, cannot be relativized without destroying its meaning completely.
>
> Gödel, "A Remark About the Relationship Between Relativity Theory and Idealistic Philosophy (1949), p. 262

Howard Stein, in his "Introductory Note" (1990) to Gödel (1949), focuses on Gödel's idealism but finds an "apparent discrepancy" here between the latter essay and Gödel (unpublished). Gödel (unpublished), he suggests, stresses the relativity of temporal relations in relativity theory to an observer, which does not, however, prevent such relations from being objective. On the other hand, he notes, Gödel (1949) emphasizes the ideal or illusory nature of time. But, in Stein's words, "to be relative is not to be illusory." This is a crucial point, and it is important to be clear on it. The resolution of the "apparent discrepancy," I would suggest, consists in the fact that for Gödel, for *time* to be relative *is* precisely for it to be an illusion. We must never forget that for Gödel the A-theoretic component of time, with its characteristic of passing by, is essential, as well as the fact that he attributes to such A-theoretic passing by a serious and absolute (i.e., nonrelative) ontological force. Recall Gödel (unpublished): "[S]omething of this kind [which lacks the character of passing by] can hardly be called time" (my interpolation). Thus, the objective but relative temporal relations of Gödel (unpublished) concern the B-theoretic structure of the Gödel universe, which Gödel believes "can hardly be called *time,*" but rather space, or space–time. Further, in a passage from

(1949) that seems almost to have been written with Stein's comment in mind, Gödel says: "[I]t may be objected that this argument only shows that the lapse of time is something relative, which does not exclude that it is something objective; whereas idealists maintain that it is something merely imagined. A relative lapse of time, however, if any meaning at all can be given to this phrase, would certainly be something entirely different from the lapse of time in the ordinary sense, which means a change in the existing. The concept of existence, however, cannot be relativized without destroying its meaning completely" (fn. 5). In a word: Some things cannot survive relativization.

That the intuitive conception of the now and of time's passage can survive in a relativistic context simply by being relativized to an observer, or frame of reference, was, for Gödel, precisely one of those misconceptions about relativity theory that prevent its revolutionary counterintuitive implications from being realized. A merely relative temporal succession, then, by Gödel's lights, is a contradiction in terms. If time were real, then by now the world-in-time would have advanced so far and no farther (recall our two diagrams at the opening of Chapter 3). This is surely what Gödel means when he says that "[t]he existence of an objective lapse of time, however, means (or, at least, is equivalent to the fact) that reality consists of an infinity of layers of the 'now' which come into existence successively" (1949, p. 262). And, of course, it is precisely the existence of such absolute, non-frame-relative, instantaneous, worldwide cuts across space–time that is denied in relativity theory. If a merely relative succession is ruled out, then (as genuinely temporal), so too is a merely subjective succession. But this last point must be handled with some care, since it involves, seemingly, establishing objective truths about the subjective. On this point, Frege (1967a) advises us to stay entirely away from the subjective, lest we try to say what cannot be said. [Frege himself, however, may not have escaped getting into trouble on this point (Hacker 1972).] One might, then, with caution, be tempted to suggest that time does exist, even in a relativistic context, but only, as it were, subjectively. For there is a sense, surely, in which even Gödel does not deny that there appears to be time. And in the case of Kant, in particular, whose idealism, as we have seen, Gödel at least in part identifies with, one might suspect that since time is, after all, the very form of inner representations, at least these representations – which Kant insists are always successive – are themselves really in time.

This temptation, however, must be resisted. It would amount to violating Kant's prohibition never to take appearances as things-in-themselves. Kant himself, moreover, would appear to have been sensitive to just this misinterpretation of his views, for he says at A38, note a, that "I can indeed say that my representations follow one another; but this is only to say that we are *conscious* of them *as in a time-sequence,* that is, in conformity with the form of inner sense. Time is not, therefore, something in itself, nor is it an objective determination inherent in things" (my emphases). For Kant, then, it would seem that I am constrained only to *represent* my representations *as* (if) in time – to *think* them, as it were, as in time. The idea that these representations of mine must somehow 'in themselves' be in time would constitute for him an impossible separation of my own representations from how I represent them to myself. In sum, *the 'appearance' of succession does not imply the (genuine) succession of appearances.* [Compare Strawson (1985, p. 39): "But what sort of a truth about ourselves is it, that we appear to ourselves in a temporal guise? Do we really so appear to ourselves or only appear to ourselves so to appear to ourselves? . . . I really do *appear* to myself temporally; but I do not really *temporally* appear to myself."] So for Kant, as for Gödel, although in some sense there 'appears' to be time, and so in this sense there is time 'relative to' our subjective sensibility, one cannot conclude from this that some things *are,* objectively speaking, in themselves in time, namely, the appearances themselves.

Although Gödel agrees with Kant, however, on the question of the ideality of time, it is important not to forget that he also strongly rejects the overgeneralization of the fruitful 'Copernican turn' into what he calls the "boundless subjectivism" of Kant's universal transcendental idealism, which brings with it the doctrine of the unknowability (by a receptive sensibility) of things (as-they-are)-in-themselves. On this point, however, there would appear to be a gap in Gödel's response to Kant in (Gödel unpublished). Kant obtains his universal idealism, let us recall, from his doctrine that the mind has a form of its own, which it imposes on its object (of thought). In virtue of this fact, the mind can never claim to know its object (as-it-is)-in-itself, though it can know (indeed, a priori) the necessary form of its imposition (the primary forms, of course, being space and time).

Gödel clearly rejects this picture, but what alternative picture does he propose to put in its place? He is silent here, but to see what a straightforward alternative would look like, we need only turn to Aristotle's account in the *De Anima,* which seems almost to have been written expressly to avoid Kant's conclusions. For Aristotle, then, the mind has no form of its own. It is a mere 'place' or 'form' of forms, and by 'containing' them, it can know them (as-they-are)-in-themselves. This view of the mind is, in fact, strongly reminiscent of the "receptacle of becoming," or space, of Plato's *Timaeus,* and we can thus characterize Aristotle's account as putting forward a conception of the mind as a kind of universal "intellectual space" of forms that has no form of its own to impose on its contents. It would be good to know if it is to a doctrine like this that Gödel, with his strong interest in Greek philosophy, would turn in rejecting the Kantian paradigm of thought and its objects. It is to be hoped that the extensive philosophical Notebooks in the *Nachlass* will shed some light on this issue.

Whatever the basis, then, of Gödel's rejection of Kant's universal idealism, it is clear that he restricted his own idealism to the doctrine of time. Gödel, after all, accepts relativity theory as, to date, the best account we have of the nature of space–time. [We set aside here the issue of the relationship of relativity theory to quantum mechanics, except to remark that for Gödel the "positivistic interpretation" of quantum mechanics is another example of an overgeneralization of Kant's 'Copernican turn': "Kant's thesis of the unknowability of the things in themselves is one example [of boundless subjectivism], another one is the prejudice that the positivistic interpretation of quantum mechanics must necessarily be the final stage of the theory" (Gödel unpublished; my interpolation).] Now, of course, a contemporary Kantian might also try to find room for relativity theory *within* his or her framework, but if we view Kant's philosophy and relativity theory from the same level of abstraction as competing overall accounts of the nature of the physical world, then they are clearly incompatible – and here Gödel sides with the latter against Kant. Note, further, that although Gödel agrees with Kant that there are certain a priori constraints on the conditions a form of being, or form of sensibility, must satisfy if it is to constitute time, unlike Kant he derives his conclusions regarding the ideality of time from what is after all an empirical physical theory – namely, Einstein's theory of relativity. One might wonder, then, how these two thinkers can derive the same results, with regard to time, from such (categorically) different sources. (We will come back to this question later.)

We are beginning to see, then, why Gödel refuses to exchange his idealism with regard to time for a more conciliatory attitude, according to which temporal notions can simply be relativized to fit in with the new relativistic framework.

The force of Gödel's position here, however, can be more fully appreciated if we look to those who have made a serious attempt to integrate into the framework of relativity theory the doctrine of temporal becoming. Among the leading proponents of such a view is Milič Čapek, who, in his imaginative study *The Philosophical Impact of Contemporary Physics* (1961), plays Heraclitus to Gödel's Parmenides. Čapek, as we saw earlier, attributes Gödel's assertion of the possibility of time travel to his excessive spatialization of time. But although it cannot be denied that there is a sense in which Gödel does treat time as spacelike, it must also be recalled that he credits this point of view on time not to himself but to Einstein. Moreover, the construction within the GTR of world models permitting time travel does not presuppose but rather (for Gödel) proves the spacelike character of time, since such models bring out consequences implicit in relativity theory itself that, for Gödel, leave no reasonable alternative to viewing time as in fact spacelike.

Unlike Sorabji (1986), Čapek does not seek a way to introduce becoming into a model like that of the Gödel universe. His response is to deny to such mathematical models any physical significance, for the a priori reason that accepting them would mean recognizing the possibility of time travel. If our suggestion before, however, was along the right lines, the question of the legitimacy – from a physical point of view – of Gödel's model, derived in the GTR, is not unrelated to the issue of his philosophical reading of relativity theory. But although Čapek raises a number of objections to Gödel's reading of this theory, his own attempt to integrate becoming with the theory of relativity runs up against what seem to be insuperable difficulties, due especially to the fact that he and Gödel agree on the ontological implications of (the passing of) time. "On each world-line," says Čapek, "my present particular 'now' separates my past from my future" (1961, p. 403). But exactly which "now" is he referring to? It can't be simply *the* now – the unique, absolute, worldwide instant of

A-theoretic temporal becoming – since, of course, there is no absolute, non-frame-relative now in the STR. Even for my own reference frame, no unique now is singled out by the STR; rather, every time, *t,* one could say, is now-relative-to-*t*. But this, the relative A-series, does not, as we have seen, discriminate between times. Further, even if I choose one time as *the* now from my own reference frame, the STR assures me that no one reference frame is privileged – so that some other observer's choice of the now from his or her reference frame has no less a claim to be *the* now. If, however, we are generous and mean to include *all* these nows as genuine, then it is unclear where in this account there is room for the *nunc fluens* – that is, for the *transition* from one time to another as the now, or the privileged ontological moment (since all nows, on this conception, are already privileged).

This difficult situation becomes even more uncomfortable when Čapek denies the relevance of the fact that his now is in the past light cone of some future observer. Merely future observers, he suggests, are not presently observable, and hence are not real. But this is true only from Čapek's perspective – here and now – not from our future observer's, there and then, and the STR precisely denies that we are allowed to take any one reference frame (or point at a reference frame), as privileged. On this point, relativity theory clearly separates itself from the classical framework. As Gödel puts it, in relativity theory, "each observer has his own set of (nows) and none of the various systems of layers can claim the prerogative of representing the objective lapse of time" (1949, p. 262). It would seem, then, that Čapek must mean that each reference frame constitutes its own separate 'river of time,' its own *nunc fluens*. And indeed, he speaks of "this plurality of contemporaneous or co-becoming temporal series" (1961, p. 364). To employ Whitrow's image, we are being asked to imagine "a complex of advancing light-cones in space–time, the track of each vertex being the world-line of a potential observer" (1980, p. 352), a complex advancing not exactly simultaneously, but rather, as Čapek says, "contemporaneously." Now Gödel does not advance the complaint of Earman (1967) that he can make no literal sense of the advancing or passing of time, but objects rather that he finds senseless the idea that "time lapses in different ways for different observers" (1949, fn. 5). For note that on the conception just entertained of multiple rivers of time, no definite sense can be attached to the question: *How much* time has flowed by *so far* (since we have no nonarbitrary point of evaluation of this question for each river or frame of reference)? As Gödel puts it, "A lapse of time, however, which is not a lapse in some definite way seems to me as absurd as a coloured object which has no definite colors" (ibid.).

Again, Čapek seems at one with Gödel that temporal becoming is an ontological notion – but then how can *reality* be 'growing' differently along (the) different lines (of the separate 'rivers of time')? "The lapse of time," as Gödel said, "in the ordinary sense . . . means a change in the existing. The concept of existence, however, cannot be relativized without destroying its meaning completely" (ibid., fn. 5). So, to sum up, there seem to be at least three challenges to those, like Čapek, who wish to include temporal becoming in a relativistic universe: (1) Can the STR recognize my set of nows as nonarbitrary (and can it even explain what it would mean for these nows to be nonarbitrary in the required sense)? (2) Do *my* nows establish a *universally valid* standard for how much time has (objectively) lapsed *to date?* (3) (What we have not discussed) Does the worldwide instant of the now reflect, as some have contended, an irretrievably conventional element (that attends all determination of distant simultaneity)?

I do not wish to state that it is impossible for these challenges to be met, but to my knowledge, no one has yet resolved all of them – not, at least, without questioning in a serious way the foundations of the STR itself. Thus, in "Time, Reality, and Relativity" (1981), Lawrence Sklar suggests that the traditional doctrine of the (exclusive) reality of the present is not decisively ruled out by the STR – but his argument (as we will see in detail later) proceeds along the lines that this theory rests, *au fond,* on an epistemological basis that constitutes a variety of verificationism. And he concludes with the strong claim that in fact we seem to find here a verificationist slide that cannot easily be put an end to, short of a kind of point solipsism. Since Gödel, however, does not argue for, but rather presupposes, the epistemological basis of the STR, nothing he has said on these matters speaks directly to this particular suggestion.

But is there not some other way to save temporal becoming? At this point it may be useful to adopt a wider field of vision, and to compare the question of the privileged position of the now, in time (or, if you will, in temporal 'space'), with that of the actual world in the modal 'space' (of all possible worlds). For just as, intuitively, one time is privileged above all others in being *now,* or the present, only one among all possible worlds is actual – namely, ours, the real world. This point of view, however, has been challenged by David Lewis (1973), who argues for the indexicality of actuality – namely, that *actual* is a property that a world has only *relative* to itself – so that all worlds emerge as equally actual-relative-to-themselves and none as the (absolutely) actual world. This is strikingly similar to the relative A-series (= B-series) picture of the now, a picture of time, we have seen, that is difficult to avoid, given the STR. I have argued, moreover, in "On Time and Actuality: The Dilemma of Privileged Position" (1986) that it is not so easy to refute Lewis's position as it has seemed to many. Nevertheless, it is certainly possible to dig one's heels in here, in opposition to Lewis, and insist, chauvinistically, that one possible world, or modal frame – namely, ours – is privileged and absolutely actual. The parallel doctrine, however, in regard to time – namely, the doctrine that one reference or inertial frame is privileged (inertial chauvinism, if you will) – is ruled out by the STR. Thus, one cannot, even by brute force (as one could with modal chauvinism), compatible with the STR, maintain that one's own reference frame (or any other) is the universally valid standard for temporal becoming. [Bearing in mind, however, our earlier quotation from Bell in regard to the Einstein versus Lorentz philosophical interpretation of the STR, it should be noted that recent experimental results, by Aspect and others, in regard to Bell's theorem (in quantum mechanics), have been held to have a special relevance to the question at hand. The matter is controversial, but Karl Popper, for one, concludes that: ". . . should the result of these experiments (especially of the experiments of Aspect) be accepted, and interpreted as establishing physical action at a distance (with infinite velocity), then these experiments would have to be regarded as *the first crucial experiments between Lorentz's and Einstein's interpretations of the formalism of special relativity*. The reason for this assertion is that the mere existence of an infinite velocity entails that of an absolute simultaneity and thereby of an absolute space" ("A Critical Note on the Greatest Days of Quantum Theory," 1984, p. 54). He adds, further, that ". . . as Newton anticipated, it may not be possible in this case to identify the inertial system that is absolutely at rest" (54).]

Now Čapek, I believe, would probably agree as to the fruitlessness of such an attempt to save becoming (leaving aside the above considerations on Bell's theorem). For example, he himself insists that the now, though ontological in force and nonarbitrary, is yet not, given the STR, absolute. No doubt, he would insist that the attempt to force a genuine now into a nonrelative worldwide instant is based on an excessively geometrical approach to the dynamic of becoming. I do not deny that this characterization fits. I do, however, wonder how Čapek intends to *coordinate* the nongeometrically recoverable dynamic of becoming with the extreme geometrization imposed by relativity theory, the epistemological basis of which he, unlike Sklar, has no wish to challenge (indeed, he seems to endorse explicitly the variety of verificationism he seems to find there). In fact, he goes further and suggests strongly that it is the very *formal-mathematical* details themselves of the STR that serve to strengthen and confirm the point of view of temporal becoming. On this point he seems to be in agreement with Stein (1968). Let us turn, then, to an examination of Stein's argument.

Stein suggests that we can recover temporal becoming from the formal structure of the STR. "The leading principle that connects this mathematical structure with the notions of 'process' and 'evolution,'" writes Stein (and *justifies* the use of our notion of '*becoming*' in relativistic space–time), "is this:

> At a space–time point *a* there can be cognizance of – or information or influence propagated from – only such events as occur at points in the past of *a*." (ibid., p. 16)

He acknowledges that this becoming cannot be considered absolute, but only *relative* to a (given) space–time point, α. "'[A]t a given time' is not a relativistically invariant notion, and the question of definiteness of truth-value [about what has already happened, of what has become], *to make sense at all* for Einstein–Minkowski space–time, has to be interpreted as meaning 'definiteness at a given space–time point (or event)' – to be vivid, 'definiteness for me now'" (ibid., pp. 14–15; my interpolation). He finds this relativizing of the answer to the question "What has become (real or determinate)?", in the STR, analogous to the relativizing of the answer, in *pre*-relativistic physics, to the question "What has become (real or determinate) *now?*"

Now, I believe that more needs to be said before we can conclude that the cases are indeed analogous. It is clear that for all space–time points, α, the STR tells us that, although what events are simultaneous with α is frame-relative, what events are *past* for α, that is, are in α's past light cone, is frame-invariant. Thus the facts that Stein has alluded to, that he believes speak to the inclusion of temporal becoming in the STR, are *geometrical-structural,* concerning the light-cone structure of space–time as it affects every space–time point, α. In what sense, however, can it be said that such geometrical facts *themselves* introduce features such as *transition, evolution,* or *becoming?* Yet Stein not only says that the STR is compatible with such notions, but even that the formal structure he alludes to "*justifies* the use of our notion of 'becoming' in relativistic space–time." With regard to the strictly geometrical relationships he alludes to, however, we seem to need only the *relative* A-series, that is, what-is-past-*for*-space–time-point-α, for *all* points α. Where, here, do we find singled out that one privileged time, *the* now, whose flux constitutes becoming? In fact, *what,* exactly *is* becoming in the picture as Stein presents it? Do we not appear to have, again, the situation Gödel speaks of, where "time lapses in different ways for different observers," of which he could make little sense, or "a lapse of time that is not a lapse in some definite way" (1949)? Again, Stein writes that "to be vivid" we can ask, in special-relativistic space–time, what has become for "*me* now?" This does seem to make this conception resemble the A-series, in which the question "What is happening?" could be paraphrased as "What is happening now, from *my* present (point of view)?" But this apparent similarity vanishes as soon as we remind ourselves that Stein is, after all, talking about *any given* space–time point α, which he has picturesquely represented as saying to itself, "definiteness for me now." What has happened here to *the* now of A-theoretic, pre-relativistic, temporal becoming?

Let us try to bring these observations on Stein and Čapek together. Two problems stand out in their attempts to include becoming in the relativistic world picture: (1) An appeal is made to the light-cone structure represented by the *geometry* of Einstein–Minkowski space–time as what *justifies* the use of such terms as "process" and "becoming" in describing a relativistic universe. (2) Although such becoming is not taken to be merely subjective, but to represent (in Gödel's words) a change in the existing, the uniqueness of the now and the existence of a (non-frame-relative) set of events simultaneous with the now have been sacrificed. In regard to (1), however, let us ask: Suppose that Einstein and Gödel are right and that the universe *is* a rigid, four-dimensional 'space'. Would it not, for all that, continue to satisfy all the geometrical-structural constraints imposed by relativity theory? Since the answer is, clearly, that it would, we want to know: (a) How can it be correct, then, to assert that it is the *geometry* of space–time itself that justifies *going beyond* such a picture of a rigid, four-dimensional space? and (b) What exactly must be *added* to the mathematical formalities of relativity theory, as it is usually presented, to render visible the extra dimension of process or becoming? In regard to (b), note that Stein would appear to follow Earman (1967) in limiting himself to a "literal-minded" representation of time (for surely nothing else is as "literal" as mathematics), though he apparently draws different conclusions, vis-à-vis becoming, from the same equations. Čapek's position, here, however, is more elusive.

In Čapek (1961, pp. 398–9), we are told that the "visual-tactual imagination" is forever inadequate to represent the dynamic of space–time becoming and that "[a]bstract mathematical constructs seem to be today the only way, not to reach, but to *represent* the structure of the transphenomenal plane." Yet, as we have seen, such abstract, geometrical reasoning is entirely consistent with a purely spacelike conception of space–time. Recall Whitrow's comment that "although Plato regarded time as an essential feature of the sensible world, he rigorously excluded it from the science of pure geometry which he associated with, and only with, the eternal world of ideal forms" (1980, p. 179). Does not a similar sentiment lie behind Einstein's comment, which forms the epigraph to Chapter 3, that though "the Now means something special for man," "this important difference does not and *cannot* occur within physics" (this from the thinker who has taken the greatest steps, in the modern era, toward the geometrization of the temporal). What is surprising, however, is that Čapek elsewhere is himself critical of the employment of classical geometry to represent time. "*[A]ll* geometrical analogies," he writes, "are intrinsically inadequate for expressing the nature of time" (ibid., p. 355). And he offers an important hint (which we will come back to later) along the same lines: "Only when succession and change are incorporated into the very foundations of geometry is there a better chance of expressing the dynamic nature of becoming" (ibid., p. 35). Clearly, then, a reconciliation is needed here, but not provided, between the view that it is the very geometrization achieved by the theory of relativity that has succeeded in "justifying" talk of temporal becoming, and the other direction of thought, put forward by Čapek, that geometry itself is "intrinsically inadequate for expressing the nature of time."

Finally, in regard to problem (2) earlier, we stand in need of a new picture of existence if we are to accept *both* that the dynamic of temporal becoming has genuine ontological implications *and* that there are multiple rivers of time, all genuinely 'flowing' and all adding to existence. Indeed, Čapek himself suggests that the very concept of reality, that is, of the whole world, is suspect in a relativistic context: "[T]o what extent does the concept of *the totality of the universe* remain meaningful after the denial of absolute simultaneity of distant events?" (1961, p. 271; my emphasis). This remark, it will be recalled, was cited in Chapter 2, in the context of a discussion of the old quarrel, initiated by Parmenides, between the concepts of reality and time. Here, indeed, Čapek would seem to be playing Heraclitus to Gödel's Parmenides. What we need to know, however, is this: (a) What are we to understand relativity theory to be a theory *of,* if not of (the totality of) the universe? (for it is surely not simply a theory of my set of nows, or the universe "to date," etc.); (b) How can one possibly have (as we do have) a *relativistic cosmology* if the very concept of "the totality of the universe" is to be abandoned? (Kant, of course, in the "Dialectic" of the *Critique of Pure*

Reason, had indeed warned us against the use of this concept. But, as we have seen, Kant's philosophical perspective was in this regard precisely at odds with the relativistic one – that is, with the perspective that Čapek himself accepts and defends.) c) How can we make sense of the genuine coming-into-being that constitutes temporal becoming if it is not unique and absolute but proceeds along parallel lines, so to speak, in different reference frames? We need an account of what Čapek calls "the *co-existence* or rather *co-becoming* or *co-fluidity* of world-tubes in relativistic space–time" (ibid., p. 377).

What problems (1) and (2) bring out, then, is a general question concerning the representation of time by a mathematical model. Without some illumination on this point, Gödel (and perhaps we should add, Einstein), on the one hand, and Stein and Čapek on the other, will seemingly be able to draw on the same formal details of relativity theory and yet reach, as we have seen, diametrically opposed conclusions. What is needed, then, is reflection on the very nature of formalization in the most general sense – including the use of mathematical models and formal systems – and representation. This we will take up in the next chapter. We conclude this discussion, however, with two revealing examples in which the pressure toward relativization, as well as toward localization, may become particularly acute.

In a well-known paper, "A Rigorous Proof of Determinism Derived from the Special Theory of Relativity" (1966), C. W. Rietdijk argued from the relativity of simultaneity in the STR to the conclusion that one can no longer consistently maintain the (A-theoretic) principle that the present alone is real. The situation he sets up is as follows. An observer, α, in inertial frame i_α, regards another observer, β, in inertial frame i_β, itself in motion relative to i_α, as simultaneous – indeed, coincident – with him, hence as real for him. Now consider an event e in α's future light cone. It is consistent with the STR that from β's viewpoint with respect to i_β, e is simultaneous with β, hence real for him. But what is real for someone who is real for me must, by transitivity, be real for me. Thus, as against the A-theoretic view of time, the future turns out to be as real as the present. Furthermore, if the future is thus, in a sense, already there, insofar as it is already fully determinate what are the truths or facts about the future, then what will be is not open. And so we must accede to temporal determinism. (Recall our earlier quote from Louis de Broglie to the effect that in space–time, past, present, and future are given "in bloc.")

How might one reply to Rietdijk if one were disinclined to abandon completely the original intuition that reality (in time) belongs only to the present? A natural response would be the one entertained by Sklar (1981) that since the transitivity of simultaneity has itself been given up in the STR, one should also be prepared to abandon the transitivity of reality. Once this is done, of course, Rietdijk's argument no longer holds. But is the price to be paid for this result acceptable? Either we would simply adopt a merely relative A-series (= B-series) or, as Sklar suggests, we could *relativize reality* to an inertial frame. Concerning the latter option, he writes: "Now the traditional philosophical exponent of these claims [namely, that only the present is real] . . . can simply accept the consequence, surprising, no doubt, but by no means inconsistent or patently absurd, of relativizing his notions of reality and determinate reality at the same time that he relativizes his concepts of simultaneity" (1974, p. 275). Now it is surely correct that no actual inconsistency arises from the suggestion that we respond to Rietdijk by simply relativizing reality itself (to an inertial frame). Yet, the following remains: (1) It becomes difficult, indeed, to recover the *motivation* for the original doctrine that the future is not (now) real if what constitutes "the" future is precisely a 'subjective' matter (since it is relative to a given inertial frame). (2) Does it make sense to relativize *reality?* (Recall again Gödel: "The concept of existence, however, cannot be relativized without destroying its meaning completely.")

Sklar (1981) broadens his response to Rietdijk by suggesting that the main result from the pressure of the relativity of simultaneity in the STR is in fact to lead us not to relativize reality, but rather to *symmetrize irreality* to what is spatially as well as temporally distant from us. Sklar has in mind the much discussed interpretation according to which the definition of distant simultaneity in the STR has an irreducibly conventional element, so that, in a strong sense, there really is no "fact of the matter" as to which (spatially) distant events are simultaneous with (hence, real for) a given event – just as there is (now) no "fact of the matter," for the A-theorist, concerning what is temporally distant from us in the future. Sklar sees this result as a consequence of the original epistemic motivation of the STR, based on the epistemic remoteness of distant events. Indeed, he claims to see no alternative basis for the STR other than the verificationism that seems to have been there from the start (in light of which the lack of a nonconventional principle of *measuring* distant simultaneity has objective significance). On this point, at least, Gödel would seem to be in agreement, for as Wang (1974, p. 12) writes, "Gödel points out that the fruitfulness of the positivistic [i.e., verificationistic] point of view in this case [i.e., the STR] is due to a very exceptional circumstance, namely the fact that the basic concept to be clarified, i.e., simultaneity, is directly observable, while generally basic entities (such as elementary particles, the forces between them, etc.) are not" (my interpolations).

For Sklar, then, the irreality of spatially remote events in the STR is consequent on the conventionality of distant simultaneity, itself a consequence of the epistemic remoteness of distant events (combined with an implicit verificationism that implies that the real coincides with the knowable or measurable). He finds here a generalization of the principle that lies behind the A-theoretic conception of the irreality of the future, founded also, he suggests, on the epistemic remoteness of the future. The question of the conventionality of distant simultaneity in the STR, however, and of its basis in a kind of verificationism, is the source of considerable disagreement among experts in the philosophy of relativity theory, and I do not propose to engage in this debate here. I limit myself to pointing out that in Michael Friedman's *Foundations of Space–Time Theories* (1983), a rather different approach is adopted to the issues raised by Sklar (see especially pp. 309–20).

What is to the point, however, is the question of whether the foundations of the pre-relativistic principle of the irreality of the future lie, as Sklar suggests, in epistemology. I would suggest, rather, that the picture is just the reverse. It is precisely because the future is not 'really there' that there arises an epistemic distance between the present and the future. In other words, the foundations of the A-theoretic intuition of the future are not epistemological but rather ontological. It is not a question of *knowing,* but rather of *being*. That is why the past, too, though determinate, is now also 'unreal,' from the point of view of the A-series. Consider, for example, our natural (pre-relativistic) attitudes toward someone's miseries in the distant past, as opposed to the present sufferings of those who happen to live in distant lands. I do not discount as unreal the sufferings of those too far away for me to reach, but I do consider as *now over* the grief of those swept away into the receding past. "Swift," W. B. Yeats writes, "has sailed into his rest. Savage indignation there cannot lacerate his breast" ("Swift's Epitaph"). Swift has sailed, for Yeats, not into a distant place, too far for us to reach him, but rather into a different mode of being – namely, nonexistence.

That past events no longer exist (and future events do not yet exist) is, I believe, the "natural ontological attitude" (to borrow, for my own purposes, a phrase from Arthur Fine), and this, rather than questions about knowledge or the flow of information (always at a finite rate in the STR), accounts for the asymmetric doctrine of the irreality of what is temporally (as opposed to spatially) remote. That people in the remote past – that is, the dead – do not exist is an intuition that, though natural, has been resisted on theoretical grounds – including considerations from the STR itself. We will come to this issue in the final chapter. What matters now, however, is that we recognize that the grounds of the irreality of the past and the future from the point of view of the A-series are ontological, *in contrast* (if Sklar [and Gödel?] are right) to the epistemic-verificationist foundations of the STR. Thus, the challenge posed by the STR to the A-theoretic view of the future remains twofold: Either (1) we *relativize* reality to an inertial frame, and thus threaten the integrity of the very concept of existence, or (2) we *symmetrize the irreality* of what is temporally remote to include what is spatially remote, introducing in the process a suspect analogy between the question of *epistemic remoteness,* which seems to some to lie behind the STR, and the concept of *existential distance,* which forms the basis of the A-theoretic attitude to the future.

However, before concluding this discussion, we should consider a way to soften the force of Gödel's conclusions, namely, by considering our pre-relativistic intuitions about present–future asymmetry to be a matter of local rather than global significance. (Michael J. White has, in fact, suggested this to me as a possible response to Gödel.) I do not believe, however, that we can find here an avenue of escape from Gödel's reasoning. Although it is certainly true that *space* can be globally curved, while remaining locally, by approximation, flat, in what sense could *time,* in the Gödel universe, be 'globally curved' yet, from a local perspective, 'flat' (i.e., preserving the felt asymmetries between present and future)? Recall that Gödel's concern throughout is not with the spacelike time of the B-series (which would permit a local–global distinction), but with the time of the A-series, with all its concomitants. But as became apparent from the two figures in Chapter 3, the very concept of a closed curve in A-theoretic time leads to antinomy. That is surely why Gödel never entertained the hypothesis that there is, after all, an "objective lapse of time," but that this "lapse" occurs *only locally.* If he objects to a merely "relative lapse," surely he will not be able to find a place in his philosophy for a merely "local lapse" of time. To suppose otherwise, to suppose, that is, that the local–global distinction applies to time, as it does to space, would be precisely to liken time – A-theoretic time – too closely to space, that is, to spatialize time. The time of the A-series, then, is incompatible with certain cosmologies, and, conversely, a proof of the soundness (or, if Gödel is right, even the possibility) of certain cosmologies would argue against the A-theoretic picture of time. Nothing similar, however, applies to *space* or to *space–time.* Here we have, then, in dramatic form, a line of reasoning that accounts for the fact, stressed earlier, that Gödel, unlike Kant, was an idealist with respect to time – but not to space (or to space–time).

The problems that arise, then, in attempts to soften the force of Gödel's results, or, more generally, his philosophical reading of relativity theory, suggest that we need to go back to the fundamental issue of how to read formal or mathematical models of time. We need to get clear, that is, on the relation of formalization, in the most general sense, to representation – that is, to the way we picture reality.

5

Formalization and representation

> Ultimately formalism in its private aspect is an expression of fear. But fear can lend us wings and armor, and formalism can penetrate where intuition falters, leading her to places where she can again come into her own.
>
> John Myhill, "On the Ontological Significance of the Löwenheim–Skolem Theorem" (1951), p. 50

A useful way of viewing the recalcitrance of relativity theory, recorded in the preceding chapter, to a reconciliation with the A-theoretic conception of an objective lapse of time is that it results from our desire both to adopt a (limited) perspective from *within* time and also to regard time itself *sub specie aeternitatis* – that is, from a position of timelessness, or eternity, in Plato's sense. That the same mathematical model of time – namely, the relativistic one – can be put to such different uses by Einstein and Gödel, on the one hand, and Stein and Čapek, on the other, suggests that we are in the presence of that "dialectic of the formal and the intuitive" that, for Wang, as we saw earlier, characterizes so much of Gödel's work. Let us recall that what Gödel achieved with his construction of a model of the Gödel universe from within the framework of the GTR was in effect a translation into a formal problem (which he then proceeded to resolve) of what was before an intuitive one (concerning the existence of an objective lapse of time in general, and in particular, within a relativistic context). Indeed, Wang (1987) speaks of the "transmutation," in Gödel's hands, of one problem into another, "cleaner," one. In fact, Gödel himself seems to have believed that Einstein had already accomplished this transmutation for the case of intuitive time. Yet, as we have seen, Einstein's translation of the problem into a formal mathematical one still left room for alternative interpretations (of the mathematical equations of the theory of relativity). To be sure, Gödel already considered some of these interpretations to be very much against the spirit of the theory; but to find something more conclusive, he derived (what we have called) a limit case for the relativistic mathematization of time – a formal result that, by his lights, no longer left any room for such non-Einsteinian interpretations.

This characterization will, I hope, help confirm the remarkable resemblance between Gödel's work on relativity theory and his better-known results on the incompleteness of formal systems. The famous Incompleteness Theorem, of course, accomplished the translation of the intuitive question: "Does there exist, even for the most powerful (consistent) axiom system, formalized along reasonable lines and designed to be adequate for number theory, an undecidable number-theoretic proposition?", into a perfectly precise formal question (which Gödel then proceeded to resolve, positively). Indeed, this translation of intuitive questions into formal ones may have been part of what so troubled Wittgenstein about the great significance attributed to Gödel's Incompleteness Theorem. "It might justly be asked," Wittgenstein remarks, "what importance Gödel's proof has for our work. For a piece of mathematics cannot solve problems of the sort that trouble *us*" (1978, VII, §22). Whatever the case may be, however, regarding Wittgenstein and the Incompleteness Theorem, it should be clear by now that Gödel's "piece of mathematics" concerning relativity theory *can* speak to the problems concerning the ideality of time that trouble *us*.

One can, indeed, go back even further and trace the source of (what might be called) "Gödel's program" to Frege's ideal of translating our intuitions of logical consequence into a formal-axiomatic system in order "to prevent anything intuitive from penetrating here unnoticed" (1970, p. 5). [However, we must be careful not to go too far and identify the precise modern conception of a formal system with Frege's. See Goldfarb (1979).] This replacing of intuition by the formal manipulation of a syntactic system leads, from Frege to Gödel to Turing, to a distinct path in the history of the concept of reasoning – the path *from the formalization of logic to the mechanization of thought.* Indeed, there would appear to be an irony here, or perhaps an instance of what Hegel called "the cunning of reason," in the fact that although Frege sharply distinguished genuine mathematical thought from the mere mechanical manipulation of formal symbols, his goal in the *Begriffschrift* was precisely to enable us to dispense with the need to invoke genuine mathematical intuition in the presentation of a proof. As he puts it in the *Grundlagen* (1980): "It is possible, of course, to operate with figures mechanically, just as it is possible to speak like a parrot: but that hardly deserves the name of thought. It only becomes possible at all after the mathematical notation has, as a result of genuine thought, been so developed that it does the thinking for us, so to speak" (p. IVe). But the very success of Frege's program pointed the way to later thinkers, like Turing, to identify mathematical thinking with mechanical computation. And Gödel himself, who writes that "[for] formal systems in the proper sense of the term, [the] characteristic property is that reasoning in them, in principle, can be completely replaced by mechanical devices" (1931, p. 195, note 70), has not been untouched by the cunning of reason. Webb (1980), for example, argues that the formal results of (the arch antimechanist) Gödel lend powerful support to mechanism. Unlike Turing, however, Gödel was not tempted to leave intuition completely out of account. [See Wang (1987) and Yourgrau (1989).] We will see later in this chapter that Frege is himself in need of an intuitive element to hold his semantics (insofar as he gave one) together, and that this intuitive element has a special relevance to time. Here, however, we can note a further point of agreement between Gödel and Frege in their interpretation of the axioms of a formal system. Whereas Hilbert and the Formal School view such axioms as implicit definitions that hold by mere stipulation within the system, and that apply to any system of elements that satisfy them, Gödel, with Frege, rejects this approach (Wang 1987, p. 247) and regards axioms as having a real, extrasystemic content, which, if the axioms are true, correctly describes a genuine (extrasystemic) reality. (This is one aspect of their ed Platonism. For more on their affinities, see Yourgrau 1989.)

In sum, the dialectic of the formal and the intuitive leads, in Gödel's work, not to the replacement of the intuitive by the formal, but rather to a translation of an intuitive problem into a formal one that, in Myhill's words, "lends us wings and armor," so that we may "penetrate where intuition [unaided] falters." The results of such an enterprise, however, have only as much plausibility and significance as does the original translation. The question of the formalization of our intuitions of time, however, presents especially acute problems in regard to the issue of representation, since the very language of the formal – namely, mathematics and logic – has been, in effect, designed expressly for the tenseless. Perhaps, then, it can help us make some progress with the problem of the geometrization of time in physics if we broaden our horizons to consider the issue of the adequacy of the representation of time in logic and semantics. Specifically, we look at an attempt to model, in current tense logic, that concomitant of A-theoretic time that regards the past as closed but the future as (now) open. After this, we come to the semantics of the crucial A-theoretic term "now" (which, we have seen, Gödel employs forcefully in his characterization of intuitive time) and discuss its resistance to accommodation within the classical semantic framework of Frege.

Aristotle, who, as we have seen, adopts an A-theoretic view of the moving now of temporal becoming, derives, in *De Interpretatione,* IX (1979), a formal result that seems to contradict a concomitant of this conception of time. Having adopted, as a basic principle of logic, the "excluded middle," p v ~p, and as a principle, in effect, of tense logic, the "necessity of the past" – namely, that what is now past is unalterable and thus (now) "necessary" – he considers propositions in the past that speak of the (then) future. But of two such opposed predictions of the future, one (by excluded middle) must obtain, hence (by necessity of the past) be *necessary.* Hence, paradoxically, the future (including any aspect that we would normally take to be contingent and thus open to chance or deliberation) turns out to be as necessary as the past. Indeed, as he says, "there is nothing to prevent someone's having said ten thousand years beforehand that this would be the case, and another's having denied it; so that whichever of the two was true to say then, will be the case of necessity" (ibid., 18^b 26). Unlike Gödel, however, Aristotle takes the philosophical implications of his formal result to be a kind of reductio, and so, applying *modus tollens,* he proceeds to reexamine the formal result itself so that the future can be seen to be, indeed, open.

Now the exact outcome of Aristotle's reappraisal has been the source of much dispute. It is not necessary, however, I believe, to enter into this controversy in great detail in order to propose the following theses, which seem clear enough from Aristotle's own words. They will help us take our bearings in the discussion that follows concerning recent attempts to model Aristotle's approach to the open future in the logic of branching time.

0. What is [true, or the case], now that it is, is (now) necessary. ["What is, necessarily is, when it is; and what is not, necessarily is not, when it is not" (ibid., 19^a 23).]
1. What is now *true* about the future is (now) *necessary.* ["For what anyone has *truly* said *would* be the case *cannot not* happen." (ibid., 19^a, 7; my emphases)]
2. *Excluded middle* holds, necessarily, for all propositions. ["Everything necessarily is or is not, and will be or will not be; but one cannot divide and say that one or the other is necessary" (ibid., 19^a, 23). That is:

   a. Nec ($pv \sim p$),

   and

   b. Nec (Fp v ~Fp),

   but not

   c. Nec (Fp) v Nec (~Fp).]

   (Read "Fp" as "p will obtain"; "Pp" as "p did obtain.")
3. *Bivalence* does *not* hold for *all* (especially future-tensed) propositions. ["Clearly, therefore, not everything is or happens of necessity: some things happen as chance has it, and of the affirmation *and* the negation, *neither* is *true* rather than the other" (ibid., 19^a, 7; my emphases).]
4. There is a *tense-logical modification* required of the "correspondence theory of truth," namely, (a) that truth comes from correspondence with the *actual* (present) facts and (b) that (the contingent part of) the future *is not now a fact.* ["[S]tatements are true according to how the *actual* things are" (ibid., 19^a, 23; my emphasis); "[I]t is not necessary that of every affirmation and opposite negation one should be true and the other false. For *what holds of things that are does not hold for things that are not* but may possibly be or not be" (ibid., 19^a, 39; my emphasis).]

   a. It follows that we must *reject* the principle that: if,

in the course of time, p obtains, then it was already true earlier to say that p would obtain.

These logical and tense-logical principles, 0 to 4, reflect the Aristotelian picture, or analogy, of time we encountered earlier, as generated by the 'moving point' of the now, a picture in which the past is always fixed and determinate but growing, whereas the future remains always (partially) indeterminate (until it arrives). Recently, an attempt has been made to provide a formal account of time along these lines, in terms of asymmetric branching models. The seminal contribution here, I believe, is Richmond Thomason's "Indeterminist Time and Truth-Value Gaps" (1970), with important papers along similar lines by John Burgess (1978), Richard Jeffrey (1980), and Michael J. White (1980). It is instructive, I think, to take a somewhat close look at the details of Thomason's formal approach. Although it is clear that the present monograph is not intended as a contribution to technical issues in the formal study of time, it remains that some important lessons can be learned by taking a look at the fine structure of one particularly elegant and suggestive formalization. (The reader who, for whatever reason, wishes to remain innocent of these formal details can simply skim the following pages and pick up the philosophical thread thereafter.)

What is the basis of Thomason's approach? We are to consider a model containing a set K, of times, ordered by a relation, $<$. This relation is not to be *linear.* So we will not have, for all times α, β, where $\alpha \neq \beta$, $\alpha < \beta$, or $\beta < \alpha$. Rather, $<$ will define a treelike structure, with the following conditions: (1) for all α, β, $\gamma \in K$; if $\beta < \alpha$ and $\gamma < \alpha$, then, where $\beta \neq \gamma$, $\beta < \gamma$ or $\gamma < \beta$; and (2) for all α, β, $\gamma \in K$; if $\alpha < \beta$ and $\beta < \gamma$, then $\alpha < \gamma$ (transitivity). The intended model, then, will have an asymmetric branchlike structure like that of Figure 3.

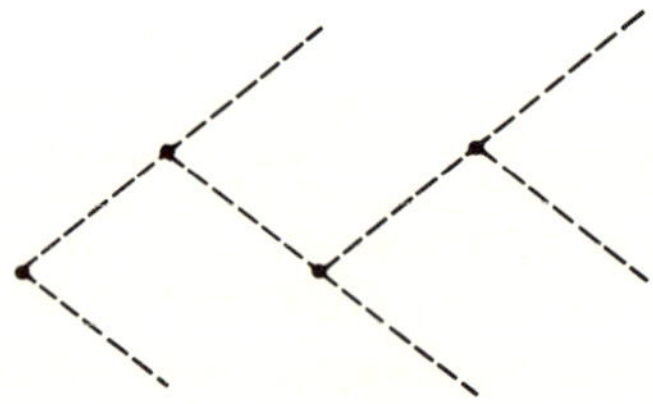

Figure 3

Notice that although, for any time α, there will be alternative possible futures, there will always be a unique path back through the past. There are many possible futures but only one past. For any time α, we define a *branch,* or *history,* through α as a subset h of K such that (1) for all α, $\beta \in h$; if $\alpha \neq \beta$, then $\alpha < \beta$ or $\beta < \alpha$; and (2) for any subset g of K such that, for all α, $\beta \in g$; if $\alpha \neq \beta$, then $\alpha < \beta$ or $\beta < \alpha$; if $h \subseteq g$, $h = g$. (Thus a branch will be a maximal linear subset of K.) We let H_α be the set of branches through α. In addition to K and $<$, a model will contain a function V, a valuation that assigns the truth values T or F to formulas. What is distinctive about such a valuation in tense logic is that it assigns a truth value to a formula only *relative to* a "point of reference." What is distinctive, in turn, about *Thomason's* tense logic is that two sorts of points of reference are employed, and there are, correspondingly, two sorts of valuations. The valuation V_α assigns a truth value to a formula relative to a time α, and V_α^h assigns a value relative to a history h through α. A formula that contains no tense operators is assigned either T or F by each valuation V_α. These valuations are thus *bivalent* over such tense-free formulas.

For the valuations V_α^h, however, the conditions are as follows:

1. a. V_α^h (A) $=$ V_α (A) if A contains no tense operators.
   b. V_α^h (FA) $=$ T if V_β^h (A) $=$ T for some $\beta \in h$ such that $\alpha < \beta$.
   c. V_α^h (FA) $=$ F otherwise.

These valuations V^h_α are thus bivalent over *all* formulas. The truth conditions of (1) seem altogether plausible. Relative to a branch h/α, "p will occur," or "Fp," should be true just in case "p" is true at some time β such that $\alpha < \beta$ – that is, at some time β that, according to branch h, is in α's future. But what of the truth value of a formula "FA" relative simply to a time α? Here we are not given a *particular future* for α; remember, too, that the picture is of a tree where α has alternative possible futures.

It is at this point that the indeterminist makes his move. The valuations V_α are designed to resemble Bas Van Fraassen's "supervaluations," and the truth conditions for complex formulas are given thus:

2. a. V_α (FA) = T iff, for all $h \in H_\alpha$, there is a $\beta \in h$ such that $\alpha < \beta$ and V^h_β (A) = T.
   b. V_α (FA) = F iff, for all $h \in H_\alpha$, there is no $\beta \in h$ such that $\alpha < \beta$ and V^h_β (A) = T.
   c. V_α (FA) is undefined otherwise.

The valuations V_α are thus *not bivalent* over all formulas. In particular, some future-tense propositions, *relative* only to a *time* α, will be truth-value-indeterminate – that is, neither true nor false. Nevertheless, as Thomason points out, the class of valid formulas on valuations V_α coincides with that for valuations V^h_α. [A formula A is *valid* just in case, for all model structures M, for all points of reference α (or h/α), V_α (A) = T for all valuations V on M.] Of special interest to the indeterminist are formulas containing the necessity, or inevitability, operator, "L." The truth conditions for formulas containing this operator are as follows:

3. a. V^h_α (LA) = T iff V^g_α (A) = T for all $g \in H_\alpha$.
   b. V^h_α (LA) = F otherwise.

("L" will be a modal operator satisfying the conditions of S5.) With this characterization of necessity, we are now in a position to evaluate the indeterminist results that Thomason derives from his logic of nonlinear time, a logic that, he says, "puts these alternatives [the possible futures] into the ontological structure of time" and succeeds "in developing a rigorous form of the traditionally popular view that "future contingent" statements can be neither true nor false" (1970, p. 265; my interpolation).

The results Thomason would like to derive are those that A. N. Prior (1967) has called "Ockhamist." The past and the present are now necessary, for it is too late to alter what is already the case. But this applies only to past and present facts that have no trace of the future in them. This is reflected in the following validity in Thomason's system: $\Vdash p \supset Lp$. Given that p contains no tense operators, it is easy to see that, for any V^h_α, if $V^h_\alpha(p) = T$, then trivially, for any $g \in H_\alpha$, $V^g_\alpha(p) = T$ [recall from (1a) that $V^h_\alpha(A) = V_\alpha(A)$ when A contains no tense operators]. This is as it should be. From the point of reference of time α, if p is true at α, then it is too late to alter this truth value – too late from the viewpoint of α or from any branch through α.

But what about $Fp \supset LFp$? A glance at (3) reveals that this will not be valid. Even if, from the perspective of some V^h_α, $V^h_\alpha(Fp) = T$, it doesn't follow that this future is necessary – for it may well be that, for some $g \in H_\alpha$, $V^g_\alpha(p) = F$. Further, $PFp \supset PLFp$ will not be valid either. For suppose that $V^h_\alpha(PFp) = T$. Then, for some $\gamma \in h$, where $\gamma < \alpha$, p is in the future. But it may well be that, for some alternative branch g through γ, p will not be in the future. Thus, $V^h_\alpha(PLFp) = F$. This fact, that it is not the case that $PFp \supset PLFp$, together with the fact that $\Vdash p \supset Lp$, appears to vindicate Thomason's claim to have produced an Ockhamist resolution of Aristotle's concerns about tense-logical determinism – that is, fatalism. The result is the more convincing in that the model structure on which Thomason bases his logic is not simply ad hoc, but rather represents a plausible construction of time. Appearances notwithstanding, however, I will argue that although Thomason has indeed produced an elegant formal reconstruction of a large part of the content of Aristotle's A-theoretic intuitions about the open future, there remain some serious questions about what the formal model has not succeeded in capturing.

A key question is whether Thomason succeeds in achieving truth about the future without admitting necessity. Note, first, that his results in fact contradict Aristotle's, if our earlier characterization of Aristotle is correct. Recall Aristotelian principle 1: What is now *true* about the future is (now) *necessary*. Has Thomason found a way out? I don't think so. The source of my concern is his condition 3, stated earlier. What he has in effect done there is to use the index of V^h_α, that is, a *branch*, in defining *truth* about the future, but then to ignore this very index in the relevant clause for *necessity* in regard to V^h_α. For he states that V^h_α (LA) = T not, as we would have expected, when A is unavoidably true *from the perspective of the relevant index*, namely, h/α, but rather, in effect, from the point of view *of time α alone*. For we are told that V^h_α (LA) = T just in case V^g_α (A) = T for all $g \in H_\alpha$. Thus, the very branch h/α that forms the index of our valuation, V^h_α, is ignored, and we are, in effect, reduced to considering the necessity of A from the point of view *merely of time α*!

But have we not slighted Thomason here? Might not someone defend clause 3 and insist that, in acknowledging that h is the way things *will* go, we do not *ipso facto* acknowledge that things *must* go that way. Indeed, Michael J. White has emphasized to me that (a) "so long as there are *multiple* branches 'to the right' of a time α, it is *possible* to distinguish, in terms of the model, between what will (actually, as a matter of fact) take place in the future relative to α and what must take place in α's future," and (b) "[t]he terms 'fixed,' 'definite,' 'determined,' 'already a fact' certainly *may* be interpreted as having a modal force, i.e., as connoting some form of necessity; but they *may* also be interpreted as connoting simply ('assertoric' – *not* 'apodictic') *actuality.* Thomason's models (as standardly interpreted) allow us to draw the distinction with respect to the future but not the past."

With regard to (a), however, we must ask what the relevance is of the multiple branches to the right of (time) α. These represent possible future histories of the world (or, if you will, possible worlds) that, *by hypothesis, will not become actual* (recall that we are considering now whether *truth* about the future entails necessity). Now a true proposition about the future, Fp, is not now necessary just in case ~Fp is now still possible. The crucial question, then, is whether the fact that there are alternative future histories from time α, including ~p, shows that ~Fp is still, at α, possible. It all turns, of course, on the relevant sense of "possible" here. But since Thomason's goal is, in effect, to formalize *Aristotle's* intuitions, we must look to his use of this modal term. Clearly, as we have seen, the modalities of necessity and possibility, as Aristotle is here employing them, are *relative* to an index – whether to a time or to a branch/history – and concern what is *unavoidable* or *unalterable, not subject to chance, or yielding to deliberation* from the perspective of the relevant index. He writes, for example that *if* "everything is and happens of necessity . . . there would be no need to deliberate or to take trouble (thinking that if we do this, this will happen, but if we do not, it will not)" (1979, 18^b, 26). What is now possible, then, in Aristotle's sense, is what can still yield to my deliberation, to my choice of action. But, by hypothesis, Fp is true, and thus ~Fp is *not* subject (now) to my control or deliberation; it is *not* now possible.

Recall, again, our background for Aristotelian principle 1: "For what anyone has *truly* said *would* be the case *cannot not* happen." The fact that ~Fp is not now possible in this *Aristotelian* sense is clearly unaffected by the fact that there remain branches/histories, or possible worlds, which by hypothesis *will* not obtain, in which ~Fp would be true. In a word, the future is not open (to my deliberation and control) if it is already true, or a fact, what will happen, tomorrow, just because there are possible histories, or possible worlds, in which the future *would* be different. These histories/branches or worlds are possible *not* in the sense that they are (now) subject to my deliberation and control, but *only* in the sense that though they are not actual, they might have been. Two senses of "possible," then, are being run together here. The fact that the Thomason model "branches out" with these possible histories only toward the future suggests here a false picture – an artifact of the formalism – concerning the multiple possibilities that seem to remain in the open future, even on the assumption that it is already a fact, a truth, what will happen in the future.

With this line of reasoning, then, we are brought back to White's remark (b). The problem here, I believe, is quite subtle. The force behind this remark is that Thomason's asymmetrically branching model shows that there are always many (possible) futures but always only one past, so that one cannot, on the basis of the model, *distinguish* past truth from past necessity. But, it would appear, the situation is otherwise in regard to the future. Here it is important to recall that we are considering the question of whether *truth* about the future entails necessity. So, *ex hypothesi,* we are now assuming that the future is as much a fact as the past. (Recall our discussion of the symmetry, with respect to facticity, of past and future times in a model of closed, circular, time – a model, we argued, incompatible with the A-series.) There are then, as we saw in our discussion of remark (a), *no more* alternative possible futures, in the *relevant* sense of "possible," than there are alternative possible pasts. In a word, the assumption of the (present) facticity of the future has the effect of collapsing all the future-directed branches into the *one* true history of the world – a *linear* history. Further, one can now ask: Given the *appropriate* sense in which these alternative branches represent possible but not actual future histories of the world, do there not also exist alternative possible but not actual past histories? Could I be myself if I had had a different past?

The Thomason model suggests that there are no such alternative possible pasts in this restricted sense of "possible"; but I think there is no *basis* for this feature of the formal model once it is given this reading. Insofar, then, as the branches were to represent the possibilities, in *Aristotle's* sense of the open future, the asymmetric, treelike structure of the model had a sound intuitive basis. For although it seems intuitive that various possible futures are now subject to my deliberation, the past is now closed and forever beyond my control. But once we assume the symmetrical facticity of past and future, the only *remaining* sense for such possible branches/histories is, as we saw, as possibilities in a different sense – mere ways-the-world-could-be-but-in-fact-won't-be. But then we *no longer* have any *intuitive* basis for the claim that in *this* sense there are many possible futures but only one possible past. We will have to bring to bear *other* considerations to settle this issue – considerations irrelevant to modeling the open future. For example, one suggestion along these lines is Kripke's (1980) thesis concerning the necessity of origin, for example, that something could not be this very desk if it had had a different historical origin – say, if it had been made from a different tree. But apart from such considerations, there is nothing in the Thomason model, given the new, restricted interpretation of the branches, imposed by the hypothesis about facts concerning the future, to help settle the question about the possibility, in this sense, of alternative pasts.

It would seem, then, that Aristotle's principle 1 – that the (present) truth of the future brings with it necessity, in the relevant sense – has not really been defeated, philosophically, by Thomason's formal model. Rather, in this regard, the branches appear only as artifacts of the formal model that can deceive us. Indeed, Aristotle's original intuition seems clear enough. If the past is now necessary because it is, all of it, a big, unalterable fact to which present truths about the past correspond, then *if* there are truths about (all of) the future, then the future, too, is a big, unalterable fact, and so it, too, is, in this sense, necessary. I detect here no sign of the modal fallacy that some have attributed to Aristotle.

Aristotle, however, as we have seen, in fact rejects the assumption that there *are* now truths concerning all of the future. Does Thomason succeed in modeling this open aspect of the future? Note, first of all, that Thomason's formal system contains, as points of reference, both individual times and branches, where the latter are, in effect, total (linear) world histories, and that, although truth-value gaps appear from the limited perspective of points in time, nothing *in* the model *prevents* the recognition of one *branch* as privileged as representing "the" truth about all of the past, present, and future. Given Aristotle's principle 1, however, which we have just defended, the existence of such a privileged branch would, of course, rule out an open future. It is true that Thomason does not in fact go so far as to recognize a privileged branch. (His system is designed, rather, to map the subtle interplay that arises between the different perspectives of points and branches.) But the fact that the existence of such a branch is not ruled out in the formal model indicates that he has not gone far enough in formalizing Aristotle's intuition of the open future. The situation here is, I would suggest, analogous to certain attempts to resolve the liar paradox through recourse to formal systems. For, just as one must be able to block the possibility of constructing, within the system, a "superliar" paradox – as in "This sentence is either false, or neither true nor false" – one must be able to block the possibility of construing one branch, from within Thomason's formal framework, as "the" total course of history (thus reintroducing temporal determinism and the closed future).

A further problem is that nothing in Thomason's formal model shows that only what is in the present – really exists, although this is a crucial feature of A-theoretic, Aristotelian time (and one that, we saw, Gödel also stressed). As far as the model goes, each individual time constitutes its own point of reference, which leaves *its* future open (from *its* perspective). But such relational facts constitute, so far, at most, a *relative* A-series. Thus, the failure of bivalence for valuations at a time, though necessary to reflect the Aristotelian picture, does not go far enough. For there to be an open future, it is not enough that if we choose to constrict our perspective to that of an individual time, the future will thereby not be reflected; rather, it is necessary that we *not be able to widen* our perspective to include (all the) facts about the future. And this negative result, we have seen, Thomason, with the availability from within his system of the (total) branches as points of reference, has not been able to establish.

Indeed, one can remark of the branching model, as we did of Stein (1968), that if the world were entirely static – reflecting no A-theoretic temporal becoming at all – it could still satisfy the formal requirements of Thomason's structure. In fact, White (1989) admits that "if such a model is to come at all close to capturing Aristotle's idea of the "developmental" character of time, we must envision the model as being viewed from a (dense and continuous) *succession* of nows along one linear (but not predetermined) branch of the model" (p. 223). This seems to me correct, but what it shows, surely, is that the formal model itself represents at best a *spacelike trace* of time. But is it not asking too much of a formal model to do more than provide such a spacelike trace? Perhaps, but then one must acknowledge that representation here must go beyond formalization and acknowledge the ineliminable role, in this context, of analogy or, perhaps, intuition. (Indeed, as we will see later, Brouwer insists that even in the case of pure mathematics, formal logic – even if modified along intuitionistic lines – is inadequate to represent genuine mathematical intuition or activity.) One might begin to wonder, then, whether the formal-mathematical method, as such, is not sufficiently out of place here to be doing a disservice to those very intuitions about time that it was intended to capture. Arthur Eddington, for example, writes, along these lines, that "[w]e have direct insight into 'becoming' which sweeps aside all symbolic knowledge as on an inferior plane" (1958, p. 97). We have already seen, however, that Einstein seems to reject such qualifications on the ability of mathematical formalism to represent time, or space–time (although the epigraph to Chapter 3 may indicate some ambivalence on his part). For now, however, we should remind ourselves that even as a formal model of time, Thomason's branching system – for all its undeniable virtues – suffers from the drawbacks to which we have been drawing attention.

The most serious limitation, however, in the model of branching time we have before us is its failure to account for Aristotelian principle 4 – that there is a tense-logical modification required of the correspondence theory of truth – although this is, arguably, the central issue concerning a correct model of the open future. Nothing in Thomason's model, however, shows why we should *accept* Principle 4, although the principle itself is nicely reflected within the system. Specifically, although $V_\beta\ (p \supset PFp) = T$ for all β, it is still the case that if (contingent) p obtains at β, it remains that, for $\alpha < \beta$, $V_\alpha\ (Fp)$ is undefined. [This is well brought out by White (1985), pp. 192–202, who dubs the former "Cicero's principle".] The question, however, is how the formal model can account for this fact. For it follows from it that someone who predicted *correctly* (as it turns out), yesterday, that it will rain today did not, for all that, *speak truly then*. Thus, even if what I say will happen, does indeed come to happen, what I said was not true! This is a glaring exception to the classical correspondence theory of truth, introduced, of course, by Aristotle, himself. But no basis for it occurs in the formal branching model of time, and so one could well object that in this regard the model merely represents a partial reflection of the complete truth about time, leaving it open that in a more comprehensive, philosophically perspicuous, systematization this as yet unexplained exception to the classical theory of truth will disappear.

What Thomason has done is to provide an elegant model of Aristotle's Principle 4a, that truth comes from correspondence with the *actual* (or *present*) facts, and of Principle 4b, that (the contingent part of) the future *is not now a fact.* (Or, at least, he has reflected these principles *from,* or *relative to,* the perspectives of the relevant indices.) Principle 4b, I believe, underlies 4a. For an A-theoretic proponent, like Aristotle, of the open future, reality literally (as Jeffrey puts it) "grows by accretion of facts." Thus, in the present moment, that is, now, the future is simply not there to correspond to. By the very nature of things, then, a future contingent proposition will, for Aristotle, always be put forward at a time that cannot "contain" all the reality required for it to correspond to, to be true. A true statement, then, for Aristotle, states how things *are,* where "are" here has essentially the *present* tense. [What, then, of the past? Aristotle seems to have in mind that the past is (now) fully determinate and that there exist, in the present, facts about the past. These facts will have to speak to properties possessed by some objects that no longer exist – for example, the dead. We come, in the final chapter, to the question of how there can be, in the present, reference to, and truths about, nonexistent objects like the dead.] The (contingent) future, then, can never confirm the past. If it rains today, it *now* becomes true to say that yesterday's prediction became true. But it remains that yesterday, before today arrived (with its rain), the prediction, lacking any actual, existent, factual correlate, was without a truth value.

How plausible, however, is this principle of Aristotle's, which lies at the heart of his conception of the open future? We will, unfortunately, find nothing to help us decide this by contemplating the formal model of branching time, for the crucial question is whether the future existence of a fact entails the (present) truth of a proposition that predicts the existence of this fact. But how does one decide an issue such as this? White (1985, pp. 195–7), who calls the truth condition according to which, what I *now* say will obtain, is *true now,* if it comes to obtain *in the future,* the "truth-value link" condition [following Dummett (1969)], argues that this truth-value link condition itself presupposes the B-theoretic conception of time (and thus, in effect, begs the question against Aristotle). I cannot agree with this assessment, however. To my mind, the truth-value link condition is based on purely *semantic* (Aristotelian!) intuitions about truth and correspondence. Since what I say (here) about New York is made true by what obtains in New York, what I say about the past is true (now) depending on how things are in the past, what I say about other possible worlds depends for its (actual) truth upon how things are in those worlds, and so on, symmetry surely suggests that what I say about the future is made true (now) by what the facts are *in the future* (entirely aside from the question of whether a fact that will obtain in the future is somehow contained in or reflected in a present metafact about the future).

In itself, I believe, the question at hand is probably unresolvable. It forces us to adopt as exclusive either a perspective from within or one from outside time, although each perspective individually is natural and unavoidable. Thus, we seem to be in the presence here of a genuine antimony. One can see, then, why Gödel was so suspicious of the notion of tensed existence and why he chose, on this point, to adopt the Kantian resolution of such antinomies – namely, idealism. For to reject the correspondence theory of truth just here is difficult indeed. If we can *think of* and *refer to* the future (which Aristotle can hardly deny), then how can we fail to think *truly* when what we say will be *will be?* Why should truth be confined to what exists *now* if the statement in question refers precisely to what will exist *in the future?* We do, after all, have the future tense, and this device seems to be able to lift us, *in thought,* out of our *existential* confinement to the present. ['Antactualists' appear to deny this. For discussion of the formalization of antactualism, and a comparison with intuitionism, see Yourgrau 1985c, pp. 556–7.] On the other hand, however, on the A-theoretic view, time is always an incomplete totality, and, as we will see in Chapter 6, there are other domains, such as the foundations of set theory, where incomplete totalities have also been invoked, and where a modification has also been suggested for the correspondence theory of truth (see Parsons 1983, p. 289). Further, if the A-theoretic picture of reality growing over time is coherent, then it is plausible to maintain that there really is no fact now, in the present, corresponding to a future contingent state of affairs, and so there is nothing in reality (as it has progressed so far) for a future-tensed proposition to correspond to. Thus, the tension between these two points of view seems to offer no hope of resolution.

Now, the question arises as to what value resides, from an ontological point of view, in purely formal models. In the dialectic of the formal and the intuitive, the formal must always be read against the background of the intuitive, rather than the reverse. Thus, even as regards the most rigorous formalism, that of metamathematics, we must always bear in mind, as Myhill (1951) says, that "metamathematics must be in the final analysis informal; for the process of discussing formalisms by means of another formalism must either terminate in a formalism which is not discussed, or be informal" (p. 42). Of particular interest to us here is the difficulty of capturing the concept of an ontologically significant privileged position, whether in time (the now) and modality (the actual world) or in semantics (the true), from within the confines of a formal model. Let us look at some examples of those features of reality that are thus formally diaphanous.

Consider, first, truth tables, especially *n*-valued tables, for arbitrary *n*. We take as an example S. C. Kleene's so-called strong table (Figure 4) for the propositional connective, "v"

Q v R

	R	t	f	u
Q	t	t	t	t
	f	t	f	u
	u	t	u	u

Figure 4

(i.e., "or") (1962, p. 334). (It is interesting to note here that the modern 3+ valued logic of Lukasiewicz originated in his attempt to formalize Aristotle's way out of the problem of temporal determinism.) Here t and f represent the designated – in our terms, privileged – values truth and falsity. From within the model, however, this can only be reflected by labeling these values the designated ones. From a formal point of view, however, this label is a mere mark without substance; all the values, considered as mere formal objects within the model, are on the same level. Indeed, this problem arises already for classical, two-valued truth tables, which are, after all, a natural reflection of Frege's initial move in taking truth and falsehood to be sentential referents – hence, to be (formal) objects. As Dummett has written, however, "it is part of the concept of truth that we aim at making true statements; and Frege's theory of truth and falsity as the references of sentences leaves this feature of the concept of truth quite out of account" ("Truth," 1967, pp. 50–1).

Indeed, the problem runs even deeper than this, for once one formalizes semantic notions one can always proceed to treat the semantic values as themselves mere formal, syntactic objects; and thus the crucial distinction between syntax and semantics threatens to disappear. As Alonzo Church puts it, "[s]emantics begins when we decide the meaning of the well-formed formulas by fixing a particular interpretation of the system. The distinction between semantics and syntax is

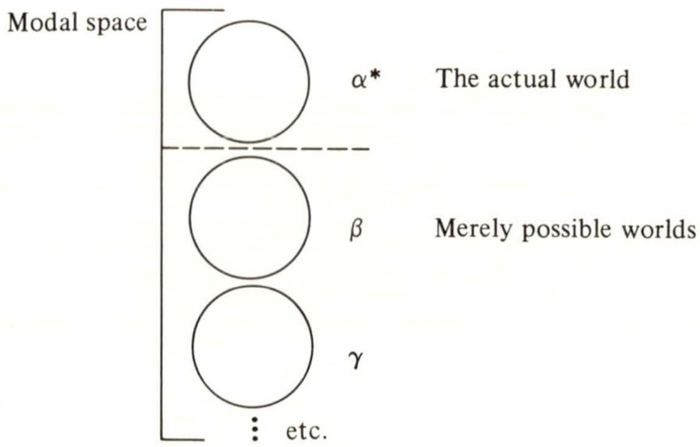

Figure 5

found in the different significance given to one particular interpretation and to its assignment of denotations and values to the well-formed formulas; but *within the domain of formal logic,* including pure syntax and pure semantics, *nothing can be said* about this different significance *except to postulate it as different*" (1956, pp. 65–6, fn. 143; my emphases).

A similar problem in regard to designated or privileged values within a formal model arises for the popular "possible-worlds" models for modal logic. Here the actual world is considered as merely one among other, merely possible, worlds in modal space, although, to be sure, the actual, or real, world is singled out by being labeled "real." Thus Saul Kripke (1963) says: "Intuitively, we look at matters thus: *K* is the set of all 'possible worlds;' *G* is the 'real world' " (p. 64). We can model modal space, then, as in Figure 5. Once again, however, it appears that what constitutes one world as privileged – here, actual or real – cannot be expressed from within the formal model (except that, in Church's words, "we postulate it as different"). Worse, as we observed earlier, one also cannot express the meta-fact that although our world, the actual world, α, happens to be actual, it need not have been so (Leibniz and modal theology to the side).

It is thus doubly misleading for Robert Stalnaker (1976) to suggest that "the actual world" has the same force as "Reality." In the first place, although Reality must obviously contain *all* the facts, including the fact that α is actual, α itself obviously cannot express this fact, for each possible world can only express, as it were, *intrasystemic* facts about the way things are *if* it obtains or is actual, whereas what is needed here is the *intersystemic* fact that among all the possible worlds, α alone does obtain – that is, a fact *about* world α, as a whole, and so not *internal* to it. [Wittgenstein calls this intersystemic fact "mystical": "It is not *how* things are in the world that is mystical, but *that* it exists" (1974, 6.44).] In addition, however, as we have just seen, α also cannot express – and, indeed, neither can the entire model in Figure 5, of modal space – the meta-fact that α might not have been actual. And a symmetrical treatment of such meta-facts leads immediately to an infinite regress: α obtains in superworld *S*, let's say, but fails to obtain in superworld *T*. But then superworld *S* itself obtains, but might not have. So, *S* obtains in supersuperworld *W*, but not in *X*, and so on.

A further question arises, moreover, concerning how seriously to take the ontological symmetry among the possible worlds suggested by the model. Are all the possible worlds genuinely *worlds* like ours? On the one side, David Lewis (1973) insists on the genuineness of this symmetry to the extent that he considers the actual world itself *not* to be privileged, absolutely speaking, in relation to the other, merely possible, worlds. "Actual," he suggests, is an indexical, like "here," so that *each* world is actual *relative* to itself. In effect, then, Lewis's modal series of worlds is akin to the *relative* A-series of times. Actuality and "nowness" (or being the present), on this conception, turn out both to be merely relative characterizations of worlds and times (see Yourgrau 1986).

On the other side, Stalnaker (1976) argues for a kind of modal chauvinism, insisting not only that ours is, absolutely speaking, *the* (only) actual world, but that the other so-called possible worlds are not really *worlds* at all, like ours, but rather complex concepts or properties. Can one steer a course between the Scylla of Lewis and the Charybdis of Stalnaker? The formal model itself is, once again, silent. Indeed, even Kripke, whose name is perhaps most closely associated with the possible-worlds model of modal logic, has had little to say on this question. Besides dismissing Lewis's modal realism (1980), he has done little more (in print) than to warn us not to take his formal-mathematical models of possible worlds too literally. But just how literally should we take them? What we need here, I would suggest, is a distinction between *being* and *existence* (and its correlate for time, *being present,* and for modality, *being actual*). We will develop this theme in Chapter 6, Section (2), but for now, we note that one can recognize the opposition between actual and possible *worlds,* between actual and possible (including past and future) *people,* and, finally, between present and future (or past) *times,* not as one between different kinds of objects, or *beings,* but rather as one concerning the *existential status* of such beings (existent vs. nonexistent, actual vs. possible, present vs. past or future) (see Yourgrau 1987b). This way, one can admit an ontological symmetry in each case concerning category of being, but also adopt, if one chooses, an attitude of chauvinism that recognizes only one member of each opposed pair as privileged (in existential status).

Let us conclude, then, this discussion of privileged position in formal models with two attempts to model time. Consider, first, the model of a world line in relativistic space–time given by a Minkowski space–time diagram (Figure 6). The dotted lines represent the paths of light rays, which thus mark out the light cones (for the diagram represents a four-dimensional space, although we have, for simplicity, collapsed the three

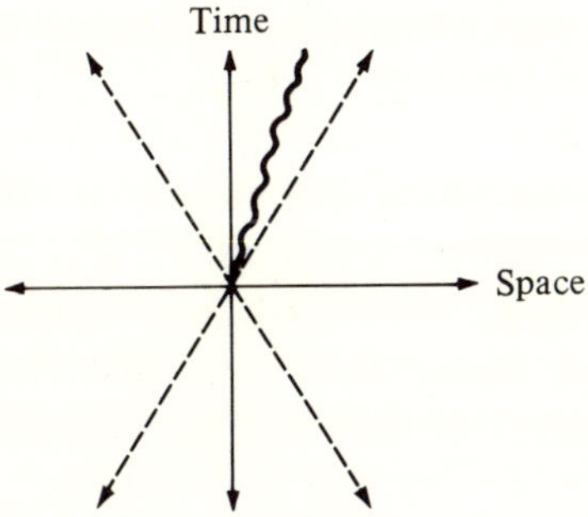

Figure 6

spatial dimensions into one), within which are the permissible paths of particles. The origin of the two axes represents the time now, but this designation is only relative to the frame provided by this particular diagram. Most important is the fact that the history through time of a particle is represented in the model as a world line. By the method of Cartesian coordinates, this world line, in turn, can be represented as a set of ordered quadruples of real numbers, and this, in turn, employing the Frege–Russell definition of numbers and the Wiener–Kuratowski definition of ordered sets, can be reduced to a large, complex, unordered *set of sets*. The formal model, that is, represents the history in time of a moving particle in a completely tenseless way. For Minkowski did not intend us to interpret a world line in a Minkowski diagram as a path a particle *moves along*. Rather, he was suggesting a *definition* of motion in terms of a *geometrical* figure – as a world *line* in a four-dimensional geometric space. [For a comparison of the problem of 'privileged position' (that is, of the now) in the context of Einstein–Minkowski space–time with the analogous problem in the metaphysics of modal logic (that is, of being the actual world) via the introduction of a combined, five-dimensional 'modal-Minkowski space–time diagram', see Yourgrau (1986).]

The approach Minkowski is offering by representing motion in time by a space–time diagram is in the spirit of one put forward by Bertrand Russell: "People used to think that when a thing . . . moves, it is in a state of motion. This is now known to be a mistake. When a body moves, all that can be said is that it is in one place at one time and in another at another" (1956, p. 1582). (It is clear that "is," here, is meant to be tenseless.) In a Minkowski diagram, then, with a world line interpreted as a definition of motion, as in the passage just given by Russell, we have an extreme example of the geometrical mode of representation. All that is left within the formal model is a spacelike trace of time. If for Plato, as we have seen, time is but a moving image of eternity, could one not say that a world line in a Minkowski space–time diagram represents a mere frozen image of time? Indeed, not only is the flux of temporal becoming, as well as the lack of privileged status for any chosen now, not represented in the Minkowski model, but, as we saw in a previous discussion, it is far from clear whether it is even *possible* to supplement these features of A-theoretic time, with the aid of intuition or imagination, given certain formal characteristics of the STR, such as the relativity of simultaneity.

In this regard, it seems, the branching model of time in Thomason's tense logic fairs better. For although it has problems similar to those of the other formal models we have considered, in regard to the difficulty of doing justice, within the model, to certain privileged moments or positions, it remains that (1) the branching, treelike structure captures an important component of the A-theoretic open future and (2) if we bracket the question of relativity theory, it seems as if we can, as White suggested, if we so choose, read into the model a dynamic conception of temporal becoming. A more accurate version of the model would then be that shown in Figure 7. Here the solid line represents the *nunc fluens,* and where its moving edge rests represents the extent that time has advanced so far. We note, then, that (a) a new diagram

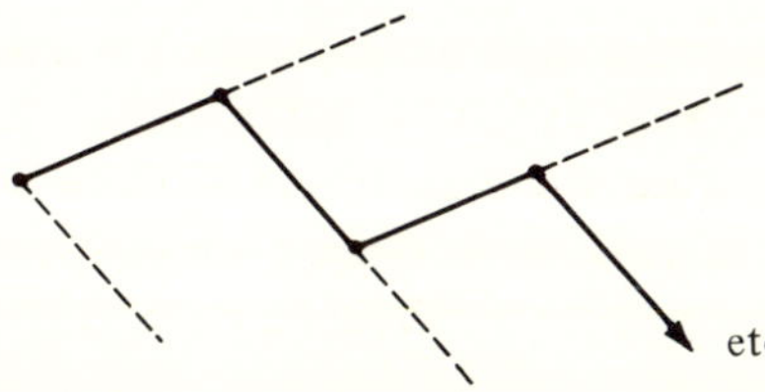

Figure 7

must be drawn to represent this feature of time, "every moment" – or else a moving diagram set up (in which case the standard, set-theoretic formulation of tense logic would need analogical supplementation); (b) the progress of time would still be represented *linearly* by the gradual, stepwise generation of a partial *line* – always (= at all times) unfolding, always incomplete; (c) the model itself still cannot be appealed to for help in deciding on the crucial questions about the open future, to wit: (1) Does truth about the future entail (temporal) determinism (= fatalism)? (2) Is a future-tense proposition now true if it corresponds to a fact in the future? and (3) Can it be ruled out that one (complete) branch is privileged and thus (already) represents the total course of history?

It should be clear, then, that in attempting to illuminate and rigorize our intuitive conceptions of privileged position, including and, perhaps, especially for the case of time, the employment of formal models cannot always be counted on to lend us, in Myhill's words, "wings and armor." It is of special note, then, that Gödel's construction of such models stands apart as being of unusual philosophical significance. For example, in regard to Church's remarks about the difficulty of maintaining, within a formalism, the significance of the distinction between syntax and semantics, we note that one of the most dramatic results of Gödel's Incompleteness Theorem is precisely to establish that one *cannot collapse* the (semantic) notion of mathematical truth into that of the (syntactic) concept of proof within a given formal system. And we recall that Gödel insisted that one of the dramatic consequences of his model of the Gödel universe, permitting time travel, was that it *ruled out* an interpretation of *t* as that time that objectively lapses. With Gödel, then, the construction of a formal model takes on a special significance for philosophy – a significance, with regard to time, we have tried to bring out in previous chapters.

We have traced, then, some of the challenges to A-theoretic time posed by formal models in physics and (tense) logic. There remains, however, a serious problem concerning the very terms we use to characterize time from an A-theoretic point of view – namely, "now" and "the present." "Now" is a member of a family of terms, including "I," "here," "this," and (more controversially) "actual," that are called "demonstratives" or "indexicals." These terms are, in a strong sense, context sensitive. "I," for example, used by me, refers to me; used by you, to you. The semantics of such demonstratives, however, including the crucial term "now," has caused serious difficulties for classical theories such as Frege's. [See, for example, the readings collected in Yourgrau (1990), which reflect also the metaphysical issues – including, specifically, the question of reality and "the now" – that must be addressed in a proper account of demonstratives.] We need, then, to see what Frege, as well as his recent critics, can teach us about "now" and, hence, indirectly, about time.

FREGE AND THE DECONTEXTUALIZATION OF THOUGHT

The moving force behind the intense concern with the philosophy of language in this century in the analytical school is the great logician Gottlob Frege – widely neglected in his own day but seminal through his influence on Wittgenstein, and thereby on Ryle and Strawson, and along different lines, on Russell and Carnap, and thereby on Quine. Frege himself, however, by training a mathematician, by inclination a philosopher, was occupied primarily in an attempt to secure the logical and epistemological foundations of arithmetic (in the broad sense), the most certain part of the most certain science. Frege's principal concern was thus with the timeless domain of pure mathematics and logic, a realm (for a Platonist like Frege) that in itself is entirely free from any contextual parameters that touch the human sphere. What distinguished Frege's conception of mathematics and logic, moreover, was his insistence that we are concerned here, not merely with a calculus for achieving certain results but also with a language for expressing truth – in other words, with a science, in the old and proper sense of the word. If, in mathematics and logic, we have to deal with language, then, in these disciplines one employs sentences and expresses thoughts and, if fortunate, attains knowledge. In this, Frege stood apart from a great many of his contemporaries, and indeed, still stands apart from many of his successors – notably, Wittgenstein ["A proposition of mathematics does not express a thought" (1974, 6.21)] and, equally notably, *not* Gödel – who in this, as in so many respects, is at one with Frege.

If, then, Frege likened arithmetic and logic to natural language, he also construed logic along the lines of arithmetic. Indeed, the title of his seminal formalization of logic reads *Begriffschrift, a Formula Language, Modeled upon that of Arithmetic, for Pure Thought.* Further, Frege introduces an analogy from another direction, likening natural language to the system of equations in arithmetic. Thus he dethrones subject and predicate from their traditional place of honor in the decomposition of the logical-grammatical form of sentences, replacing them with the mathematical categories of argument and function. Truth and falsity, then, emerge as the (truth) values of sentential functions (which he calls "concepts"), given, as arguments, the objects referred to by singular terms. [We saw earlier that Dummett (1967) has expressed reservations about this construction of truth as a truth *value* – that is, as the value of a function, hence, as an object, in just the sense as are the arguments of a function. [For a reconstruction of a formal proof that, in a Fregean context, the designation of a sentence must be its truth value, see Gödel (1944). For an argument that this formalization could not be used by Frege to establish his thesis, see Yourgrau (1987).] For Frege, then, no line separates (natural) language, logic, and arithmetic insofar as each provides, in a suitable manner, for the systematic expression of thought.

In regard to logic, in particular, Frege insists that it concerns truth in the most general way – indeed, that its end or purpose is to expose the very laws of truth (-preservation, in inference). He makes, further, a bold transcendental assumption, in the spirit of Kant, that the truth bearer, the thought, must be true in itself, and so not merely true *relative* to context – whether we include as context time, place, or speaker. (Recall Kant's insistence that if ethics is not to be an illusion, something must be capable of being good in itself – namely, the will.) In this, as in his conception of logic as a science as opposed to an *organon,* he takes exception to Aristotle, who views the truth bearer as precisely context *relative.* ["Suppose, for example, that the statement that somebody is sitting is true; after he has got up *this same statement* will be false" (Aristotle 1979, 4a 22; my emphasis).] In contrast, Frege writes: "But are there not thoughts which are true today but false in six months time? . . . [F]or example, that the tree there is covered with green leaves. . . . No, for it is not the same thought at all" (1967a, p. 37).

If, then, mathematics is a science, which pretends to knowledge, we must demand a justification for its claims. With Kant, Frege sharply distinguishes the (causal-psychological) origin of our beliefs from a 'deduction' or demonstration of legitimacy. Only the latter concerns him. Further, with Plato, he requires of knowledge of a domain that one know the objects whose structure is being investigated – that one recognize, as it were, the fish one has trapped with his net. But how does one go about justifying a vast enterprise like mathematics? Here Frege takes a leaf from Descartes's book and seeks out the least foundation on which the enterprise as a whole can be found to rest. He divides the task into the question of geometry, on the one hand, and arithmetic (including number theory and analysis), on the other. The former he takes to have been settled already by Kant, as resting on the basis of a (synthetic) a priori intuition of the pure form of space. The latter task is made propitious by the fact of the gradual arithmetization of mathematics, culminating in the nineteenth-century arithmetization of analysis. (Descartes's arithmetization of geometry is, of course, of no use to the geometric Kantian, Frege.) The effect of such arithmetization is the gradual shift in mathematics from the qualitative to the quantitative. More specifically, the multiplicity of categories of mathematical being – natural numbers, negative numbers, rationals, irrationals, imaginaries, and so on – is replaced by diverse manners of grouping (via sets, or classes, or extensions) a single kind of object – namely, the natural numbers. Whence Kronecker's famous remark that "God created the integers, and man has made the rest." [For René Thom, however, "this maxim, spoken by the algebraist Kronecker, reveals more about his past as a banker who grew rich through monetary speculation than about his philosophical insight. There is hardly any doubt that from a psychological and, for the writer, ontological point of view, the geometric continuum is the primordial entity." (1971, p. 74) He adds, further, that "perhaps, even in mathematics, quality subsists, and resists all reduction to sets." (p. 77) We will return, later, to the issue of the reduction of intuitive mathematical ideas to the set-theoretic.]

The possibility thus opens up of actually being able to solve, in brief compass, the problem of foundations (beyond geometry), if only the following specific tasks can be accomplished: (1) The natural numbers themselves must be defined

– and not just distinguished from other categories of mathematical being. (For the language of mathematics is, we recall, construed here as referring to a definite reality – and we need to know *which*. Thus, as Frege says, we require that it be settled, for example, whether 0 = the moon!) (2) We must explain the possibility of cognitive access to such objects as the natural numbers. For Frege notes that the paradigmatic form arithmetic knowledge takes is the equation, which, by his analogy with natural language, he takes to be a statement of *identity,* flanked on both sides by singular designations – numerals (i.e., proper names) or definite descriptions (since "2 + 3" becomes "*the* sum of 2 and 3") – which thus designate *objects*. But Kant had insisted that our only method of access to genuine objects was by intuition, and Frege, unlike Kant (as well as Aristotle and Brouwer), denies that we have (synthetic) a priori knowledge of the numbers via our intuition of temporal succession. Thus he must either relinquish his claim that numbers are genuine objects or provide for some other, nonintuitive, mode of access. (3) The permissible rules of deductive inference by which we derive the theorems of arithmetic must be made explicit and justified, as well as the deductive reasoning required by arithmetized domains like analysis. (4) The basic laws of arithmetic must be exposed and justified or else derived from yet more basic laws. (5) An account must be provided of those groups or sets by which the arithmetization of mathematics is accomplished.

Now what is remarkable is that Frege had good reason to believe he had actually accomplished all five tasks required for foundations. (1) In his *Grundlagen* and *Grundgesetze* he provides explicit definitions of the natural numbers as logical objects. This is made possible by the fact that his logic is in fact higher order, containing both "concepts" and "extensions" (his version of the "groups" or "sets" needed for this enterprise), exploited in his definitions. Thus, what Kant never dreamed of, pure logic shows itself to contain its own objects. [That such Frege-style definitions of the natural numbers in terms of extensions, or, in Russell's case, sets, are unavoidably arbitrary, and thus unsuccessful, is argued for in Benaceraff (1965). Yourgrau (1985a) suggests that the problem with Frege's definitions lies deeper than this.]

(2) Since, on Frege's account, numbers turn out to be logical objects, our cognitive access to them will be via the logical faculty itself, which Frege takes to be one of the three "Sources of Knowledge of Mathematics and Natural Sciences" (1979) (the other two being the geometrical source of knowledge and sense perception). Frege, however, never speaks, to my knowledge, of the faculty of logic delivering logical *intuitions*. Indeed, he has very little to say about this faculty, since, for him, its proper place of investigation lies not in logic (which investigates only the relationships between *truths*), but in psychology. Again and again, he seems to characterize what is clearly epistemological as psychological (and thus to dismiss it).

(3) In *Begriffschrift* (1970), Frege had accomplished the formalization of quantification logic (whose first-order part Gödel proved complete), which for the first time permitted the systematization of the principles of deductive reasoning required for the nineteenth-century rigorization and arithmetization of mathematics. Further, Frege claimed to have reduced all peculiarly arithmetic modes of reasoning – such as mathematical induction – to the purely logical (made possible by his definitions in task 1). (Indeed, this was only to be expected, given his insistence on the universal applicability of arithmetic – a trait it clearly shares with logic.)

(4) In *Grundgesetze* he attempted to derive the basic laws of arithmetic from the basic laws of logic. [For the relations between Frege's formulation of the basic laws of arithmetic and those of Dedekind and Peano, see Gillies (1982).] Here, again, Gödel provided the relevant meta-mathematical proof – only this time, surprisingly, in the negative, demonstrating via the arithmetization of meta-mathematics that any reasonable, formal, axiom system for arithmetic must be (if consis-

tent) incomplete (and this for the most certain part of the most certain science).

(5) In *Grundgesetze,* Frege provided a formal account of the groups or extensions required for the completion of his project. In particular, his (infamous) Axiom V established the relationship between a concept and the extension (or class) it determines. Unfortunately, for a third time, a problem arises, since Bertrand Russell showed that Axiom V, in the context of *Grundgesetze,* leads to inconsistency. Indeed, the problem here is especially interesting, since it consists, at least in part, in the desire we have already met with several times in this study to adopt simultaneously a perspective within and outside of a totality. In Frege's case, a quantifier was permitted to range over, as a member of its domain, the extension of a predicate that falls within the scope of that quantifier. [For a discussion of Russell's solution to this problem via his "vicious circle" principle, see Gödel (1944).]

Having accomplished, then, as he believed, all five tasks required to secure for arithmetic, and so for the rest of mathematics, a firm foundation, Frege addressed himself to a final task that even then remained unfulfilled. For it will be recalled that arithmetic truths take the fundamental form of identities. "Identity, however," Frege writes, "gives rise to challenging questions which are not altogether easy to answer." With these opening lines of "On Sense and Reference" (1892), a new chapter commences in the philosophy of language, or (as Frege sees it), the philosophy of thought. But what are these challenging questions identity gives rise to? A true identity seems somehow to be about, to concern, *two* things, but also, in another sense, to be about *one*. But as Wittgenstein, a close student of Frege's thought, puts it, "to say of *two* things that they are identical is nonsense, and to say of *one* thing that it is identical with itself is to say nothing at all" (1974, 5.5303). Such difficulties led Wittgenstein to abandon the concept of identity entirely. They led Frege in *Begriffschrift* to interpret a statement of identity, "a = b," to be, in terms of its logical grammar, about the signs "a" and "b" themselves. Now, however, in "On Sense and Reference," he finds this artificial translation unsatisfactory. To begin with, it introduces an unjustified change of subject matter. "Mark Twain is Samuel Clemens" is clearly about Mark Twain the man, not "Mark Twain" the name. Further, that two signs – that is, ink marks or sounds – happen to be used (somewhere, by someone) to designate the same object is of little interest, since "[n]obody can be forbidden to use any arbitrarily producible event or object as a sign for something" (1892, p. 57). But identities like that just given can be of considerable interest.

Reversing our tack, however, produces no improvement. If we take identity to concern not the signs but rather what they designate or represent, then all true identities turn out to be trivialities (of the form that a given object is itself). But, again, many identities are so far from being trivialities that in fact they represent great advances in our understanding – as in the famous example of Hesperus being Posphorus. Further, if all identities are trivialities – and indeed, on this account, are knowable a priori – the same would hold of *all* propositions, since all statements can be reexpressed as identities. [Thus, for an arbitrary statement, "Fa," we have:

"$Fa \equiv a = \iota x\,(x = a \cdot Fx)$."

Indeed, Gödel relies on this equivalence in his reformulation of Frege's argument that the designation of a sentence is its truth value (Gödel 1944, p. 122, Note 5).] One can see why Wittgenstein was inclined to give up identity entirely rather than to seek its elusive explication. His position, here, moreover, has special relevance for mathematics, where equations are essentially employed. "It is the essential characteristic of mathematical method that it employs equations. For it is because of this method that every proposition of mathematics must go without saying" (1974, 6.2341). Yet, as we have just seen, any arbitrary statement can be recast as an identity. So, along these lines alone, the response to be made is surely that if the propositions of mathematics do not express thoughts, or "go without saying," the same applies to *all other* propositions! Frege, however, chooses a different path. He accepts the apparent logical form of the equations of arithmetic as genuine and seeks a deeper analysis of identity.

The problem, he now discovers, is that we have not yet recognized the need to split up the meaning – or "content," as he put it in *Begriffschrift* – of a term or sentence into two components. Words do not simply *represent* objects, but rather *present* them in one way as opposed to another. This mode of presentation by which we introduce items into our thought is a matter for cognition. Meaning is here a question of "*cognitive* value." This is then to be sharply distinguished from the question of the *truth* value of a sentence, or the issue of designative value, or reference. We can then lay it down: Whereof there are *two* in an identity is material for *cognition;* whereof there is *one* concerns *truth and reference.* Frege calls this mode of presentation, which raises a mere sign to the level of a representation, a "sense," and he says that the sense of a sentence is a "thought." A proper name, then, designates its referent via its sense, and a sentence designates its referent – a truth value – via the thought it expresses. The subject of logic, then, is thoughts, not sentences, and the cognitive value of mathematics as a branch of knowledge lies on the level of sense, not reference.

What exactly are senses, however? Frege proceeds to derive their nature from their functional characterization. An empirical object as such, and also a mental-psychological object, does not, he insists, point beyond itself. A sentence, after all, is a mere picture in words, but even a picture that obviously resembles its object – say, a realistic drawing of a pipe – does not thereby represent it. Something else must be added, some principle of projection, to connect this one object with another. Without this we face, one might say, the "Magritte problem." I have in mind here Magritte's famous painting

"Ceci n'est pas une pipe," in which a lifelike drawing (as of) a pipe has painted under it those very words: *Ceci n'est pas une pipe.* If we restrict ourselves to the domain of images ("ideas," in Frege's terms), the painting can never be seen to point beyond itself unambiguously to (represent) a pipe. For as much as the drawing points (by resemblance, via an *assumed* principle of projection) to a pipe, the painted title erases (by its assumed meaning, as another kind of painting or image) the customary or assumed meaning of this very drawing.

By cutting loose from their safe, ordinary contexts such typical vehicles of representation as a realistic drawing and a simple sentence, and by letting each challenge the other's referential credentials, Magritte has succeeded in undermining our presupposition that somehow the mere painted images themselves succeed in referring. For here the images seem to be quarreling with each other, and *confined to the level of images* alone, no resolution of this quarrel is in view. Thus, considered as a mere painted object, including as a part a painting of a sentence (for a sentence just *is* a painting of a sentence), the picture never succeeds in pointing beyond itself. A mere picture, as such, always stands in need of an interpretation (Frege 1967a, p. 18). But where does this leave us? Is anything true in itself, or does everything stand in need of an interpretation? Frege's answer is clear. Unless there is something that is literally true, true in itself, there can be no truth at all, and rational thought threatens to disappear. There must, then, be such a truth bearer, true in itself. Frege calls it the "thought." The thought, for Frege, is intrinsically articulate: It says what it means. He concludes that it would have to occupy neither the empirical nor the mental-psychological realm (since the same considerations would apply to mental images). Rather, "[a] third realm must be recognized" (1967a, p. 29). [Indeed, Frege would surely not be surprised that the resolute commitment of W. V. Quine *not* to follow Frege down the path to this "third realm" (of senses, or, in modern terminology, "intensions") has led to the so-called indeterminacy of translation and the abandonment of analyticity. See Katz (1988).]

Senses, then, including thoughts, turn out, from an *ontological* point of view, to occupy a timeless, abstract, Platonic third realm, also occupied by concepts and numbers. Functionally speaking, they satisfy a variety of roles. They serve to present an object in a definite manner in our thoughts and so to account for the cognitive value, or 'information', contained in our sentences. By determining specific objects as referents, moreover, they also help determine the truth value of sentences. Finally, they provide the indirect or oblique referents in sentential contexts where what is at issue is not (ordinary or nonoblique) truth and reference. For example, in so-called propositional attitude contexts, such as "I suspect that Mark Twain is Samuel Clemens," the sentential component "Mark Twain is Samuel Clemens" refers in this context not to its ordinary reference – a truth value – but to its ordinary sense – namely, the thought that Mark Twain is Samuel Clemens. Finally, from a logical point of view, we note that Frege makes no attempt to formalize his account of senses. [This has been done, however, by Church (1951, 1973), though serious problems remain even with these formulations.]

What is of most concern here, however, is the role of context in the expression of thought via language. Although "[t]houghts are independent of our thinking" and "whoever thinks it encounters it in the same way, as the same thought" (Frege 1979, p. 127), and indeed, "[i]t is thus of the essence of a thought to be non-temporal and non-spatial" (ibid., p. 135), it remains that words alone do not always suffice for the expression of a thought. It is exactly here that context comes into play. "If a time indication is needed by the present tense one must *know when* the sentence was uttered to apprehend the thought correctly. Therefore the *time* of utterance is part of the *expression* of the thought" (Frege 1967a, p. 24; my emphases). Frege is sensitive to the fact that mere words may well stand in need of contextual supplementation in order for a complete thought to be expressed or apprehended. The thought itself cannot always be put completely into words. Nevertheless, one can attribute to Frege here a "principle of exportation": The pragmatic contribution of *context* can always be completely exported into (not necessarily *words,* but) semantic *thought content.* [See Yourgrau (1983) for an epistemic formulation of such a principle.] In this sense, Frege has brought about a *de-contextualization of thought.* Indeed, when the relevant contextual parameter is time, one may have to change one's words, over time, to keep expressing the same thought. "If someone wants to say the same today as he expressed yesterday using the word 'today,' he must replace this word with 'yesterday' " (ibid., p. 24).

This example of Frege's, however, has inspired Gareth Evans to offer a very different account of the relation of Fregean thought to context demonstratively indicated by words like "today" and "yesterday." On Evans's proposal, the Fregean thought expressed by "Today is fine" today is a *tensed* thought, which one can grasp tomorrow by using the word "yesterday." "Might Frege not have had in mind an idea of a thought the grasp of which, on a later day, requires just as specific a way of thinking of a day as does its grasp on an earlier day – namely as the preceding day?" ("Understanding Demonstratives" 1985). Further, "the grasp, at [time] t, of a thought of the kind suggested by the passage from Frege, a *dynamic* Fregean thought, requires a subject to possess at t an ability to keep track of a particular object over time" (ibid., p. 87).

As an interpretation of Frege, however, Evans's reading appears to miss the mark doubly. In the first place, from our previous discussion, it should be clear enough that there is no such thing as a *dynamic* Fregean thought. Recall only that "[i]t is of the essence of a thought to be non-temporal and non-spatial" and that "[w]hoever thinks it [a thought] encounters it in the same way, as the same thought." In the second place,

Evans's 'dynamic Fregean thoughts' are not even genuinely *dynamic*. For Aristotle, for example, the thought expressed by "Today is fine," though true today, may be false tomorrow. For Evans, however, this same thought remains true tomorrow, no matter what the weather, only tomorrow it is expressed by the words "Yesterday was fine."

Assuming, then, that our reading of Frege is not undermined by Evans, it emerges that Frege draws a sharp line between thought content and the process of thinking. Not only is it true, as we saw earlier, that for him "thoughts are independent of our thinking," but thinking itself, as a mental-psychological process, would appear to have, for Frege, no internal relation to thought (content). But how can this be? Isn't thinking, after all, in the sense of "having thoughts," a natural mental process? Well, is it? Perhaps it is simply a highly evolved biological process, like digestion? But my digestion cannot contradict yours. Must the process of thinking contain its objects? Must I "absorb" this tree in order to think it? Senses and "concepts" (in Frege's sense), no more than trees, are absorbed into the mental process. Thus, in the Preface to the *Grundlagen* (1980), Frege clearly places *both* concepts and objects on the side of the objective, while mental ideas belong to the subjective. The problem here is that the categories proper to thought – namely, truth, implication, validity, and so on – do not seem to have any internal relation to the empirical or psychological realm, which knows only motion, contiguity, resemblance, causation, and so on. Hence the clash between Fregean semantics and the recent causal-genetic theory of reference of Kripke, Donnellan, Kaplan, Putnam, and others (addressed in detail in Yourgrau 1987c).

Here we should recall our discussion of Magritte's painting and the inability from within the empirical realm, which includes the relation of resemblance, to provide a basis for an account of the representational function of Fregean sense. Indeed, Frege goes to great pains, in "The Thought," to distinguish the relation "thought → truth value" from the relation "picture → object (depicted)." The strain this produces between the perspective within nature and the point of view of pure thought (*sub specie aeternitatis*) makes it always tempting to adopt a kind of "dual aspect" view of the world – the phenomenal aspect and the transcendental – in the manner of Kant. Frege, it would appear, has nothing new to offer here to relieve this strain, and shows no inclination to adopt, under such pressure, a merely forced synthesis. Indeed, if anything, he tends to draw attention to the irreconcilability of the two perspectives. His integrity here, however, puts enormous stress on his ability to avoid outright paradox, as when he writes that "still the grasping of this law [of gravity] is a mental process! Yes, indeed, but it is a process which takes place *on the very confines of the mental* and which for that reason cannot be completely understood from a purely psychological standpoint" (1979, p. 145; my interpolation and emphasis). It is, I take it, a mystery what could constitute being on "the very confines" of the mental realm (short of a dangerous spatial-geometric metaphor).

Frege, however, further destabilizes an already unstable position by insisting that thought, in itself immaterial, can make itself apparent to us only by wearing a 'garment' of (sense-perceptible) words. "I have to content myself," he writes, "with presenting the reader with a thought, in itself immaterial, dressed in sensible linguistic form" (1967a, p. 26, fn. 1). The choice of metaphor, surely, is not a happy one, and renders his task more difficult and mysterious than it need be. Just as my body does not *clothe* but rather *embodies* my soul, it would be more suggestive for Frege to say not that language "dresses up" thought, but rather that it embodies it. The thought is the soul of the words, what gives them life and meaning. What is clear, however, is that Frege resists the tendency, since the "linguistic turn," to remove the tension here and to simply identify the thought with the words (in context). For Frege, we must *resign* ourselves to the fact that we can only, given the kind of beings we are, encounter thought in an alien medium. Nevertheless, he insists that we can, via words, grasp thoughts. Indeed, we read in "The Thought" that each of the three realms – the sensible, the mental, and the third realm – is accessible to us by the appropriate mode of cognition. We *sense* a physical object, *have* a (mental) idea, and *grasp* a thought.

What is peculiar to the latter mode of cognition, he insists, is the lack of reciprocal action so typical of the natural world. To grasp a thought is very different from holding on to, say, a hammer. "How different the process of handing over a hammer is from the communication of a thought. The hammer . . . is gripped, it undergoes pressure and on account of this its density . . . is changed in places. There is nothing of all this with a thought" (1967a, p. 38). Like an Aristotelian unmoved mover and (I would suggest) a Platonic Form, a Fregean thought moves by being apprehended, though remaining itself unmoved. "How does a thought act? By being apprehended and taken to be true. . . . There is lacking here something we observe throughout the order of nature: reciprocal action" (ibid., p. 38). For Frege, if thinking, grasping a thought, somehow disturbed it, or put it into motion, then the objectivity of thoughts would be lost, as would their constancy and stability. In a passage strongly reminiscent of the *Theaetetus,* he writes that "[i]f everything were in continual flux, and nothing maintained itself fixed for all time, there would no longer be any possibility of getting to know anything about the world" (1980, VII). Thus, grasping a thought, knowing something, must not imply affecting the thought in any essential way. In a word, "[i]f knowing is to be acting on something, it follows that what is known must be acted upon by it, and so, on this showing, *reality when it is being known* by the act of knowledge must, in so far as it is known, be *changed*" (Plato, *The Sophist*, 1973, 248e; my emphases). On this point, clearly, Frege is at one with Plato.

A difference emerges, however, as soon as it becomes a

question of that being that is not changeless, of the flux of time and human experience, of 'reality in motion'. Here Plato comes closer to saving the phenomena. "But tell me, in heaven's name, are we really to be so easily convinced that *change,* life, soul, understanding have no place in *that which is perfectly real. . .* ?" (1973, p. 249; my emphases) Frege's extreme separation of thinking from (what is) thought (in his strict sense), his preoccupation with the logic of mathematics – or 'reality at rest' – leads him, in spite of strong contrary inclinations, to a kind of cognitive cul-de-sac, where we somehow cannot *put into thought* what we can nevertheless *grasp in experience,* and so to neglect the very form of that other reality, the reality in motion, namely, *time.* For recall that Frege recognizes, in "The Thought," three categories of cognitive access, each suited to an appropriate category of being: We *have* (mental) ideas, *experience* sensible objects, and *grasp* thoughts. Yet the official doctrine he puts forward is that all reference is only via senses (the components of thoughts); all thinking that attains to *truth* and *knowledge* is constituted *only* by grasping (eternal, changeless, third realm, Fregean) *thoughts.* In a word, it seems that we cannot, for Frege, succeed in having objective thoughts that are directly about what we directly experience – either the ideas we have or the sensible objects we see or touch. It is more, then, than the fact that we cannot always succeed completely in putting what we experience *into words.* This Frege acknowledges even for a complete *thought.* Rather, it is a question of what can be *put into a thought,* and *so known* – in the sense of knowledge that implies truth.

There is, then, in Frege's own account of reference and thought a tension that demands resolution. For even his own thesis in "The Thought" concerning the three modes of cognition, each proper to its own ontological realm, is rendered incomprehensible given his official doctrine of no reference without *sense,* no thinking without *thought* (in his sense of this term). Thus his thesis implies that we have, but do not see, an idea of the color red. Yet *this* thought cannot contain this experienced idea, but only a sense, a conceptual correlate (in the intuitive, not Frege's, sense) that determines this idea. So, the direct deliverances of experience are, as such, invisible to Fregean conceptual thought. But the problem is even worse than this. For if the only means of "introducing an item into our thought" (to borrow a phrase from Strawson 1963) is by *designating* it via a *sense,* then *this* sense, too, must be introduced into our thought by itself being designated by a new $sense_1$. But $sense_1$ must then itself be introduced by a new $sense_2$, and we are landed in a (vicious) infinite regress. To avoid it, however, we need only bring in the other half of Frege's theory, the half that acknowledges that we can *grasp* thoughts and other senses – as opposed to only *designating* them via other senses. But once we recognize this, nothing prevents us from continuing in the same direction, and thus from also acknowledging that we can introduce into our thought the deliverances of immediate experience, simply by having such experiences (ideas, in Frege's sense), without going through the conceptual intermediary of a designating or referring *sense* – which, moreover, must, by its very nature, miss precisely what is peculiar and proper to the world of direct experience of ideas. For the mode of presentation of the experience of red *as* that very experience is essential to it. (Recall our earlier discussion of Sklar on the secondarization of the phenomena as science – or, in general, abstract, conceptual thought – advances, and also Kripke's important reminder that when it is a question of the mental experiences themselves, their mode of presentation – for example, of a pain *as* painful – is essential.) Thus, if we are not to change the subject when attempting to think about the deliverances of experience and to capture, instead, a mere sense that happens to designate or refer to that experience, we must acknowledge having an experience as a legitimate mode of presentation of a *referent,* along with designation via *sense.*

Thus, I lay it down that Frege cannot but acknowledge that

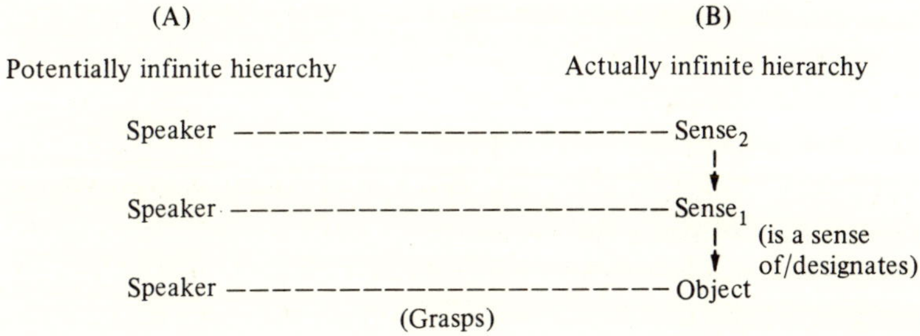

Figure 8

all three direct modes of cognitive access to the three realms – having, sense experience, and grasping – are as legitimate modes of presentation of our referents as is designating by a sense (which, alone, is recognized by the official doctrine). For the case of grasping senses, we can represent the situation diagrammatically in Figure 8. Column B represents a metaphysical hierarchy of senses in the realm of pure thought. Column A, however, represents an epistemological (Frege would say, mental-psychological) domain ("hierarchy" is not quite the right term), which shows us how we grasp the elements of Column B. [For details, see Yourgrau 1982, including the question of "grasping a grasping" and the issue of the hierarchy of iterated "oblique" or "indirect" senses in propositional attitude contexts, made famous by Church (1951) and considered by Davidson (1968, p. 99), a reductio of Frege's theory.] It is not necessary here to go into the fine points of the nature of this relation of grasping, but it is important to note that it, together with the cognate relations of having and of sense experience, are *cognitive* relations. An analogy would thus be Russell's notion of *acquaintance,* not the causal-genetic relation employed by the causal theory of reference, alluded to earlier. [For details on why these Fregean relations must be cognitive, see Yourgrau (1982, 1987c). For a comparison of *Russell,* acquaintance, and the Theory of

Descriptions with *Frege,* grasping, and the theory of sense and reference, see Yourgrau 1985b.]

It should be clear that, depending upon whether the object at the base of Column B in Figure 8 is an idea or an empirical object, the diagram with represent either the relation of having an idea to designating/denoting such an idea by a sense, or the relation of the sense experience of an empirical object to designating/denoting such an object via a sense. In each case, Column A represents a kind of direct, nondescriptive, nonconceptual mode of cognitive access to the relevant ontological realm. But when the realm in question is temporal becoming – given, perhaps, with Kant, through the "form of inner sense" – may not Column A represent something more pertinent to our present concerns, namely, a kind of *tensed access to a tensed reality?* Indeed, in naming it "Column A," I have tried to suggest just such a kinship with the tensed reality of A-series time. One would here, then, bring Frege into line with Plato, whose attempt to save the phenomena of cognition, time, and change in *The Sophist* we have already alluded to, and which Iris Murdoch has well characterized by saying that Plato "here conjures up *a moving knower to follow a moving known*" (1978, p. 49; my emphasis).

If this suggestion is along the right lines, then it becomes understandable why the only form of reality directly visible to Fregean thought is tenseless – namely, because our access to tensed reality, reality in motion, is not via the conceptual hierarchy of pure thought, represented by Column B, but rather through the direct, nonconceptual, epistemic (or, for Frege, mental-psychological) channels of cognitive access represented by column A. For as Bertrand Russell has put it in "On the Experience of Time" (1915): "Thus *introspection* is necessary in order to understand the meaning of 'past,' because the only cases in which this relation is immediately given are cases in which one term is the subject [of experience]. Thus 'past,' like 'present,' is a notion derived from *psychology*" (p. 192; my interpolation and emphases). [And recall Čapek

(1961, p. 372) on Whitehead and the "direct awareness of time."]

Even if our proposed reconstruction of Frege is successful, however, it does not, of course, establish the objective existence of the tensed reality of the A-series, but rather creates some *semantic breathing room* for the recognition of A-theoretic facts, should they exist – in particular, should they survive the challenges posed by Gödel and Einstein. But without such a reconstruction, and assuming, as we have suggested, that Evans's dynamic Fregean thoughts are the result of a misinterpretation, it would be a consequence already of the *semantics* of temporal discourse – without any help from Einstein and Gödel – that the A-series is an illusion. And, indeed, this result *has* been claimed on behalf of semantics alone – not, however, by an unreconstructed Fregean, but rather by the severest of critics of Fregean semantics. How did this come about?

Coordinate with the criticism of Frege advanced by the causal-genetic theory of proper names, there is a critique centered on the study of demonstratives, inspired by the non-Fregean theory of such terms put forward by David Kaplan. [See, for example, "Dthat" (1978) and "Thoughts on Demonstratives" (1990).] Bringing an analysis like Kaplan's to bear on Frege's remarks on demonstratives, like "now," John Perry, in "Frege on Demonstratives (1977), argued that Frege could not adequately account for both the *dynamic* aspect of such referential devices (each time you use "now," a different time is indicated) and the apparent *constancy* of meaning that underlies such shifts in reference (for in some sense, "now" is being used in the same way across these changes in reference). For Perry, Frege, using only the theoretical devices of sense and reference, cannot explain this apparent constancy through diversity. He puts forward, therefore, a theory of demonstratives more in the spirit of Kaplan. He distinguishes the *content* that is expressed by a demonstrative statement from the *expression* of such thoughts. The content, for example, of "Today is fine," asserted on day *t,* is the ordered pair consisting of the actual day, *t,* and the property of being fine: ⟨*t,* (being fine)⟩. Thought *content* gives us the *information* carried by such assertions and grounds *truth value.* The expression of such content, by contrast, is determined by what Perry calls the "role" of the demonstrative. The role of "today," for example, is to refer, when used on day *t,* to *t.* The role of a demonstrative explains our psychological states, hence our actions, when employing such devices. Thus, when you *entertain* the role of "today," on day *t,* in asserting "Today is fine," you refer to *t* and adopt the psychological attitude to day *t* that inclines you to put away your umbrella. In contrast, entertaining the role of "yesterday," on day $t + 1$, to assert "Yesterday was fine," issues in the same content, ⟨*t,* (being fine)⟩, but expresses a different psychological attitude: You don't yet know whether or not to bring along your umbrella when going out.

Now, though neither Perry nor Kaplan drew the consequences of such a semantic account of demonstratives for the question of A- vs. B-theoretic time, D. H. Mellor, in *Real Time* (1981), applying Perry's analysis to the crucial temporal demonstratives "now" and "today," argued that the B-series alone constitutes 'real time,' the A-series being a kind of semantic hallucination produced by an incorrect account of the crucial terms. Indeed, I would suggest that a strict employment of the Kaplan–Perry theory of demonstratives, or, as they are sometimes called, "indexicals," must inevitably issue in a kind spatialization, in the sense introduced earlier, of the relevant domain. Thus, the indexical account of "actual" advocated by Lewis, which we alluded to earlier, is a kind of spatialization of actuality based, in effect, on a Kaplan–Perry approach to this term. Since, for Lewis, the complete truth is given by the fact that "actual," used at possible world *w*, refers to *w*, it follows that no single possible world is singled out as privileged – as *the* actual world. Rather, all worlds turn out to be actual *relative* to themselves. Similarly, Mellor argues that the A-theoretic notion of *the* now is based on a (semantic) misconception, since *all* times are now *relative* to themselves. Thus Mellor, along semantic lines, finds no reason to recognize anything but the relative A-series – that is, the B-series. Unlike Gödel, however, he does not conclude that time is ideal, but rather, as we have seen, that time, though real, is characterized exclusively by the B-series.

We will not repeat Gödel's reasoning that, in effect, the nonexistence of the A-series would imply the ideality of time. Rather, I wish to focus on the adequacy of Perry's theory itself as an account of "now," as well as on Mellor's pragmatic account of the illusory intuitiveness of the A-series. The crucial question for Perry's account concerns the entertaining of a role. What exactly *is* a role? A role, it would seem, is a semantic rule. The role for "now," then, would be of the form: "For all times *t* and speakers *s,* if *s* employs 'now' correctly at *t,* he or she refers to *t.*" Now a rule is no good unless you can use it, but if you attempt to employ this semantic rule, it becomes apparent that, in grasping it, you get a handle not on any *particular* time, but only on a *universal conditional* on times and speakers. The problem, then, is that to actually put this rule into effect, you must *instantiate* the universal quantifier. Yet, to accomplish this, you must already have a particular time, *t,* in mind. *But how do you get to have it in mind?* By describing-denoting it, for example, with a B-theoretic time specification – say, July 31, 1990, 8:00 P.M. – that is, in effect, via a Fregean sense that designates its object timelessly? But this would be a throwback to Frege and is, accordingly, solidly rejected by Perry. How about simply taking *t* to be the present moment, that is, now? This, of course, would be circular, for it is the role = semantic rule itself that was to explicate our use of the term "now." One is soon forced to the conclusion that Perry has mistaken a necessary constraint on a mode of presentation, or mode of designation, of a time with "now" – that, whatever this mode of presentation may be, it must be such that, employed at time *t,* it determines *t* – for the mode of presentation itself.

I conclude, then, that Perry has failed to provide an account of temporal demonstratives alternative to the Fregean, and therefore, that the door remains open, from a semantic point of view, to a recognition of the A-series if a reconstructed Fregean theory, like the one we gave earlier, is sound. But if Perry's account fails of its purpose, then so must Mellor's employment of it as a semantic critique of A-theoretic time. Mellor also has, however, a positive account of the relative A-series that he puts forward as an attempt to soften the paradoxes that might otherwise seem to be consequent for the human experience of lived time, if an objective, nonrelative 'now' proves an illusion.

An important aspect of the seeming significance of the present, the now, is that present experiences have a different value in our lives than do those in the past or future. Thus, in a famous essay, "Thank Goodness That's Over" (1959), A. N. Prior argues forcefully that it is difficult to account for our relief when, for example, a *present* pain is *now over,* if tense and the nonrelative now of the A-series are mere illusions. (Recall again Einstein's statement that "the *experience* of the Now means something special *for man.*") Mellor (1981) suggests that what is crucial in the kind of example Prior raises is the relationship (B-theoretic) between the time of the experience and that of the utterance of a token of the appropriate sentence. Thus, "Thank goodness that's over" is uttered immediately after the painful experience has ended, and "I wish this pain would end" is uttered simultaneously with the pain.

Is this, however, a plausible explanation of the phenomena Prior adduces? Recall that without the A-series time is spatialized, so that positions in time bear no ontological significance (and this applies to all events, including the mental). Thus, all my pains, past, present, and future, are, by this account, equally real. But why, then, would it be such a fundamental feature of human existence that we attribute great significance to the fact that a pain (or a death – as we will see in the next chapter) is simultaneous (or not) with the utterance of a particular sentence of ours? After all, although the pain is simultaneous with this utterance, it is not for that reason more real than any other pain of mine, past or future. It merely has a particular location in (B-theoretic) time, just as, say, a pain in my foot has a particular location in space. One might, of course, have a foot fetish and pay special heed to pains in that location, and similarly, one might have, I suppose, a temporal fetish and be especially concerned with pains occurring at certain temporal locations, such as being simultaneous with certain sentences I utter. But we seem in any case not to have come to grips with the essence of the phenomenon of lived time Prior has drawn attention to – the human significance of the present moment.

Thus it seems that if the A-series is an illusion, it will not be so easy to avoid paradoxical implications. Gödel, moreover, seems rather to insist on these paradoxical consequences than to attempt to soften or sidestep them, and Einstein, as we have seen, senses here a dilemma. It remains, however, that even if the challenges to the objectivity of A-theoretic time, and hence to time as such, from the side of tense logic and semantics are not conclusive, the problems raised by Gödel, from the point of view of physics, are considerable. Rather than press on, however, in further consideration of the challenges to the reality of time, I suggest that we consider next what happens if we follow Gödel and adopt an idealistic conception. In short, what exactly must we give up if we abandon A-theoretic time? Too high a price, here, may incline us rather to a *modus tollens,* against Gödel, rather than a costly *modus ponens.*

6

Being and time

> [The Philosopher must be] [l]ike a child begging for "both," he must declare that reality or the sum of things is both at once – all that is unchangeable and all that is in change.
>
> Plato, *The Sophist* (1973), 249[d]

> Heidegger singles out time as more fundamental than space. . . . [But] [e]ven as an account of how time looks to man this phenomenology is defective, since it rules out the possibility of the presence of the eternal or genuinely trans-temporal within the very structure of *to phainomenon*. . . . Plato comes closer to saving the phenomena.
>
> Stanley Rosen, "Heidegger's Interpretation of Plato" (1988), p. 146

If, with Gödel, we maintain the ideality of time – that successive time that unfolds into an open future – we abandon not only our overtly temporal intuitions (which Gödel recognizes), but also certain otherwise clear, natural, and intuitive distinctions of a mathematical, as well as of a more general, human, significance. There is, to begin with, the traditional association of time with potential infinity.

TIME AND INFINITY

It has been the custom in foundations of mathematics, at least since Aristotle, to distinguish *actual,* or completed, infinity from infinity that is merely *potential,* or always incomplete. Potential infinity, however, held the field among mathematicians (the other variety being restricted by the theologians to God) until the work of Georg Cantor in the late nineteenth century. Indeed, as late as 1831, Gauss himself could write that "I protest above all against the use of an infinite quantity as a *completed* one, which in mathematics is never allowed" (in Dauben 1979, p. 120). A quantity is potentially infinite if it increases without limit or if it goes on forever. With the latter formulation we see time enter the picture, and it must clearly be successive, forever incomplete, A-theoretic time. Thus Kant writes that "the infinity of a series consists in the fact that it can never be completed through successive synthesis. It thus follows that it is impossible for an infinite world series to have passed away" (1965, p. 396). And Cantor writes that "the potential infinite is mostly witnessed where one has an undetermined, *variable finite* quantity which either increases beyond all limits (here we can take as an example that so-called time which is counted from a definite initial moment)" (in Hallett 1984, p. 12). Finally, Mary Tiles sums up the matter well, leaving no doubt that it is A-theoretic time that is here at issue: "Potential infinity is opposed to the sort of completeness required of the universe, just because the universe consists of what is, of what is *actual,* whilst the potentially infinite is never fully actual but always in the process of becoming. *It is an inescapably temporal notion*" (1989b, p. 28; second emphasis mine).

One could thus say that the set-theoretic Platonist, Cantor, by introducing actual infinity (indeed, as we will see later, as the primary concept), was in effect breaking the historical tie between time and infinity, as Plato had separated time from geometry. And it is clear that Cantorian infinity is opposed specifically to A-theoretic time. For suppose, employing the latter concept, that time is infinite – that is, goes on forever. Could it have gone on *even longer?* This obviously makes no sense for this sort of infinity. Yet, distinctive of Cantorian infinity is that it is extendable: for any infinite cardinal, α, Cantor proves that $2^\alpha > \alpha$.

Thus, if Gödel is to be believed that (A-theoretic) time is an illusion, it follows that, at most, B-theoretic structures characterize the universe. But nothing essentially incomplete survives in the B-series. Thus one can put forward a characteristically Gödelian disjunction: *Either time is not ideal, or the only permissible mathematical infinity is the actual.* For if time is ideal, then, assuming, with Kant, that it is the very form of inner sense, it follows that not only are there no potential infinities in the natural world, but even the mathematical imagination will be unable to fashion such a concept as objective. This result about the epistemology of infinity, if sound, would clearly threaten the approach of intuitionistically inspired philosophers, like Dummett, for whom "it is quite literally true that we can arrive at the notion of infinity in no other way than by considering a process of generation or construction which will never be completed" (1978b, p. 56). (Of course, one must be careful, as we have already stressed, not to conflate Gödel's idealism with Kant's. Thus, Gödel's idealism applies only to time, *not* to relativistic space–time, *or* to mathematics – especially with regard to Cantorian, set-theoretic, actual infinity.) That such results in *pure* mathematics seem to follow from considerations on the nature of time derived from the *empirical* theory of relativity may give us pause. Yet, in retrospect, this was to be expected. For the internal relation between A-theoretic time and potential infinity is an indication that here mathematics is not being cut free from its human context – that it is not being de-contextualized – and thus that the thinking *process* is here not disconnected from (mathematical) *thought.* (This becomes especially clear in the conception not just of potential infinity, but of mathematics in general, advocated by L. E. J. Brouwer, to be discussed later.)

Is there any direct mathematical evidence that counts against the second disjunct in our Gödelian proposal – and so favors the existence of potential infinity? What, after all, is a genuine witness for potential infinity? Jonathan Lear, in "Aristotelian Infinity," suggests that a temporal process of dividing, though incomplete at any finite stage, can be a witness to

Aristotelian potential infinity. "[A]n actual process of division, which terminates after finitely many divisions, is all that a witness to the existence of the potential infinite could consist in. While such a process is occurring, one can say that the infinite is actually coming to be" (1980, pp. 191–3). Lear goes on to suggest a translation of a crucial line in Aristotle, *Metaphysics,* VIII.6 (1048 b, 14–17): "The division not being exhausted ensures that this activity (of dividing) *is,* potentially, but not that the infinite exists separately" (ibid., p. 193; my emphasis). Lear glosses this: "On this reading it is precisely because there are possible divisions that will remain unactualized that the line is potentially infinite" (ibid., p. 193).

Now I cannot see how, for Aristotle or Lear, any actual temporal process of only *finitely* many stages can itself bear witness to the existence of the *infinite.* We need a firmer basis for, as Gödel puts it (Wang 1987), "the first infinite jump" (which Gödel sees as the decisive leap of idealization here). Indeed, Lear seems to be in some sense aware of this when he adds that it is "because there are possible divisions that will remain unactualized" that such a finite process can serve as a witness to infinity. But it remains that (1) there will always, for *all* finite possible divisions, remain new possible divisions to be made, and (2) we must have some *epistemic access* to this modal fact – namely, that we *could* always make (in some merely possible world) more divisions than we ever in fact (i.e., in this strictly finite, actual world) will. How, then, can "an actual process of division, which terminates after *finitely* many divisions" provide one with the needed epistemic access to this modal fact?

Further, as Frege wrote, in a review of Cantor, "many mathematicians and philosophers only acknowledge the potential infinite. Cantor now succeeds in showing that this infinite *presupposes* the actual infinite, that the 'receding limit' *must have an infinite path,* if it really is to recede ever and ever further" (1979, pp. 68–9; my emphasis). Indeed, it is instructive to compare Frege's remarks here with our earlier discussion of branching time and with the question of whether the ever-diminishing open future also presupposes an actually infinite path, or branch, along which this diminishing takes place. In any case, we must have another form of access to such an infinite path than the (A-theoretic) temporal one (which can bring us, at most, an incomplete or potential totality). Thus, if Cantor and Frege are correct, the advocates of potential infinity must resolve not only the previous epistemic problem but also the question of our access to actual infinity (the very question, no doubt, they were trying to avoid).

Whether or not, however, Lear or Aristotle can adduce a witness to potential infinity, and thus, if our disjunction is valid, cast doubt on the thesis of the ideality of time, it remains that Gödel's commitment to Cantorian set theory and actual infinity is untouched by any results about time. For Gödel combines his Platonism in mathematics with a strong commitment to mathematical intuition, which he refuses to identify with a Kantian–Hilbertian concrete space or time intuition. (See, for example, Gödel 1964 and Yourgrau 1989, p. 395.) In a word, Gödel harks back to a kind of purely noetic or *intellectual* intuition, reminiscent rather of Plato than of Kant. [It would be instructive to compare this conception with Charles Parsons's (1980) more Kantian account of mathematical intuition.] Indeed, Gödel is explicit that such intuitions appear precisely to lift us out of the realm of the merely empirical into a kind of contact – an *analogue* of sense perception – with the mathematical world. But if Gödel is correct about time, and if we are then not to occupy the untenable position of failing to recognize any category of mathematical infinity, it would seem that he has found surprising new evidence concerning his, to date, far from influential conception of mathematical intuition.

Epistemological issues to the side, however, one should note two ontological aspects of A-theoretic time brought to the foreground by the issue of potential infinity: its directionality and its incompleteness. Recall Cantor's remarks in which he relates the potential infinite to "that so-called time which is counted *from* a definite initial moment." A potentially infinite series, then, has a definite origin and a direction from this origin. And this direction (from, one might say, the potential to the actual) is not reversible. A putatively potentially infinite series whose direction is toward the past, with a present member of the series as the origin, would, of course, be rather an actually infinite totality (since, for the present moment to have been reached, the former moments – all infinitely many of them – must have already elapsed, or become actual). Thus we have Kant's famous argument, in the First Antinomy, that the infinity of time can only be a feature of its future, not of its past (since, as we have seen, he recognized only the potentially infinite). Here, it would seem, mathematics and cosmology meet again, the direction of influence this time, however, coming from permissible mathematics to permissible cosmology. On this point, a vast debate has raged, over 'Kant vs. Cantor'. (See, for example, Huby 1971, Whitrow 1978, and Craig 1979.)

With regard to the infinite past, I would suggest that, in order to keep a clear view of the logical picture, we must keep in mind the following relationships: (1) (From the mathematical point of view) There is no strictly mathematical reason not to maintain the existence of both (a) potential and (b) actual infinities, so long as the epistemological basis of each is sound – for the case of (a), A-theoretic temporal intuition, for (b), if Gödel is correct, intellectual intuition. Thus the cases here are independent of each other, and are dependent on cosmological considerations only to the extent that these may speak to the validity of these epistemological bases – as we have suggested that Gödel's views on the ideality of (A-theoretic) time cast doubt on the legitimacy of the appropriate intuition of the potentially infinite. (2) (From the cosmological point of view) If time is, indeed, A-theoretic, then the past (unlike the future) cannot be infinite. But this fact would have no relevance, if Gödel's epistemology is sound, to the question of Cantor's set-theoretic conception of actual infinity. It would, however, be highly relevant to the question of the interpretation of relativity theory, if Gödel's arguments in this regard are sound.

But the connection of the directionality of time to questions in the foundations of mathematics involves more than the issue of the irreversibility of the direction of potential infinity. Let us distinguish the *order* of a series from its *direction*. Following Russell (1903, p. 210), let us say that a series of elements is *ordered* by a relation, R, if and only if, for each x, y ($x \neq y$), either xRy or yRx (i.e., R is connected), and R is asymmetric (hence, irreflexive) and transitive. The *direction* of relation R, then, is from x to y if xRy. Further, one speaks of *the* direction of a *series*, which is thus unique and 'privileged.' Thus, for example, the series of natural numbers (construed Platonistically) is ordered by the relation 'less than,' but also by 'greater than.' Unless one direction, then, is singled out as privileged – for example, by the A-series of time – there will be no such thing as *the* direction of this series. Indeed, Hans Reichenbach remarks (1958, p. 139) that the STR seems to require only that time be *ordered* in the preceding sense. And, I would suggest, recent attempts to analyze the direction of time in terms of entropy are in fact more relevant to the question of *order* than to that of *direction*.

Now let us recall, with this distinction in mind, two recent reductions of intuitive mathematical notions to strictly logical/set-theoretic ideas. (1) Intuitively, in the ordered pair $\langle x, y\rangle$, x comes before y in terms of direction. But modern standards of rigor (i.e., of formalization) demand an explicit definition of such pairs in terms of (unordered) sets. The criterion of adequacy of such a reduction is usually given as $\langle x, y\rangle = \langle w, z\rangle$ if and only if $x = w$ and $y = z$. This, then, is satisfied by the now standard Wiener–Kuratowski definition: $\langle x, y\rangle = \{\{x\}, \{x, y\}\}$. Indeed, W. V. Quine calls this a "philosophical paradigm," since it illustrates so well that "the question 'what is an ordered pair?', is dissolved by showing how we can dispense with ordered pairs in any problematic sense in favor of certain clearer notions" (1979, p. 260). It is clear that what Quine means by "clearer notions" are the logical/set-theoretic replacements for intuitive (here, temporal) ideas.

In terms of our earlier distinction, however, we observe that the Wiener–Kuratowski definition of an ordered pair preserves at most order, not direction. For an equally valid definition along the lines of this kind of reduction would be $\langle x, y\rangle = \{\{y\}, \{y, x\}\}$. The intuition, then, that in the ordered pair, $\langle x, y\rangle$, x comes first or precedes y has now been lost. And there is an additional problem here, analogous to that raised by Benacerraf (1965), for Frege's set-theoretic definitions of the natural numbers (more strictly, definitions in terms of extensions of concepts): to wit, that although an ordered pair is supposed to *be* an unordered set, there seems to be no nonarbitrary way to determine just which *particular* unordered set the ordered pair is supposed to be identical with. (See Dipert 1982 and Kitcher 1978.) [One should also note here that Cantor himself, the seminal figure in (quasi-spatial, as Gödel says) set theory, took the notion of ordinal, or ordered sequence, as basic (see Hallett 1984, p. 8). If, then, the preceding remarks are to the point, we would seem to have intimations of the temporal in the very heart of the 'space' of Cantor's sets.]

Along the same lines as this kind of problem – that is, preservation, through reduction to logic/set theory, of order but not direction – lies, I would suggest, Frege's purely logical reconstruction of the natural number series. His definition of the successor of a natural number is, as he insisted, of a purely logical/set-theoretic nature, with nothing of the temporal remaining. Yet just this temporal (A-theoretic) element in the notion of the successor of a natural number was insisted on by Aristotle, Kant, and Brouwer. The replacement, however, of this unformalized, intuitive-mathematical conception by what Quine calls the "clearer notions" of logic/set theory leads to the preservation of the order of the intuitive, preformalized, natural number sequence, but not of its direction. And it is, of course, no accident that Frege, contra Aristotle, Kant, and Brouwer, considers this series to be actually infinite. If he, too, recognized it as only potentially infinite, then he would not have been able to disregard the question of the direction of this series – since, as we have seen, the direction of a potentially infinite sequence is not reversible (on pain of relinquishing its status as potential). [And, of course, the (actual) infinite enters Frege's formalization of logic and mathematics in the form of polyadic quantification. What Aristotle and Kant could only represent in terms of iteration in time, Frege can express in terms of quantifiers. See Friedman (1990, pp. 233–4).]

It would seem, then, that the direction of modern work in foundations of mathematics toward formalization and thus toward the detemporalization of notions that seem otherwise to have an intuitive, A-theoretic basis may be leading us to lose sight of features of the mathematical world more visible to those less inclined to divorce the process of mathematical thinking from its thought content (in Frege's sense) – that is, to those who have no wish to decontextualize mathematical thinking. (See Wang 1961.) Indeed, it seems that, often, when attempts have been made to banish all traces of time from foundations of mathematics, something temporal manages, nevertheless, to slip back in unnoticed. We conclude this discussion with an example of this phenomenon of special relevance to our concerns.

Another familiar mathematical distinction that threatens to disappear along with (A-theoretic) time is that between *complete* and *incomplete* totalities. The potential infinity of natural numbers constitutes one such incomplete totality (closely related, clearly, to the incompleteness of the A-theoretic open future). But there are others. Cantor himself pointed out that the collection of all (in themselves complete) transfinite ordinals does not constitute a complete totality – namely, a set.

Indeed, he identifies completed mathematical totalities with sets and appears to regard mathematics as the science of such completed totalities. He writes, for example: "When . . . the totality of elements of a multiplicity *can* be thought without contradiction as 'being together,' so that their collection into 'one thing' is possible, I call it a *consistent multiplicity* or a *set*" (Hallett 1984, p. 34). But, he adds, "[a] collection can be so constituted that the assumption of a 'unification' of *all* its elements into a whole leads to contradiction, so that it is impossible to conceive of the collection as a unity, as a 'completed object.' Such collections I call *absolute infinite* or *inconsistent* collections" (Dauben 1979, p. 245). The problem, however, is that the meta-theory of set theory will clearly have to have recourse to just such "inconsistent collections," since, for example, the universe of *all* sets is not a set. (The proof is due to Cantor himself, who shows easily that although the "universal set" would have to contain *all* sets, its own power set would be still larger, since, if the cardinal number of the universal set is α, the cardinal number of its power set is 2^{α} – and we have already seen that $2^{\alpha} > \alpha$.) The meta-theory must employ, then, collections that are not sets – that is, incomplete totalities, known since Gödel as "proper classes." The question, then, is whether the meta-theory of Cantor's paradigmatically Platonist, nontemporal theory of completed totalities, or sets, can itself be intuitively grounded without invoking those temporal (A-theoretic) ideas that we have found to lie behind other incomplete totalities, such as potential infinity and the open future.

A case in point is Charles Parsons's discussion of the intuitive basis of set theory in "What is the Iterative Conception of Set?" (1983c). At issue is the so-called iterative conception of set, favored by Gödel, who writes that "this concept of set, however, according to which a set is something attainable from the integers (or some other well-defined objects) by iterated application of the operation 'set of', not something obtained by dividing the totality of all existing things into two categories, has never led to any antinomy whatsoever . . ." (1964, pp. 258–9). A set, on this conception, presupposes, for its existence, the 'prior' existence of its members (from which it is 'obtained'). As is perhaps already apparent, however, this 'genetic' conception may suggest to the imagination ideas of time and process (a set 'emerging' from the collecting together of 'prior' members), and Parsons argues that in his attempt to provide an intuitive picture of this conception Wang (1974), chapter 6, comes too close to actually employing such notions. After all, "set theory," as Parsons says, "is the very paradigm of a *platonistic* theory" (273). Seeking, then, to avoid the (on the surface) temporal and the (more deeply) 'idealistic' strain, he finds (268) in Wang's intuitive model, Parsons urges that "what we need to do is to replace the language of time and activity by the more bloodless language of potentiality and actuality" (293). Drawing on Cantor's idea that the elements of certain ("inconsistent," as Cantor puts it) multiplicities (such as the 'universe' of all sets) cannot all "exist together" and thus do not constitute a set, Parsons suggests that elements that exist together *can* constitute a set (293). Further, for Parsons, this holds for any multiplicity of objects that can "exist together," where the latter appears to mean, or to imply, that they exist together in some possible world. ("I assume," he says, "that in any possible world {z: Fz} is the set of existent z such that Fz in that world" (p. 286, note 25). Clearly, moreover, no one possible world can be "maximal," or closed under set formation, since "there is for any world *i* possibly a set of all objects existing in *i* that are not elements of themselves, and that set cannot exist in *i*" (Parsons, 1983d, p. 330). Finally, he invokes modal notions again in his account of "inconsistent multiplicities," themselves. The "universe of all sets," he suggests, cannot presuppose the prior existence of all its members, since there is no such "prior existence." Rather, it is, for him, a so-called proper class, something essentially potential. A proper class, moreover, for Parsons, is something "predicative," kin not to an *extension* (of a predicate) – which has closer historical and conceptual ties to *sets* – but to an *intension,* "if we think of an intension in the traditional way as a meaning entertained by, and in some sense constructed by, the mind . . ." (1983c, p. 285).

Now, I cannot hope in the present context to do full justice to the richness of Parsons's account (which, moreover, seems to have developed from 1983c to 1983d), but I do want to advance some considerations that suggest, to my mind, that here, too, temporal (A-theoretic) ideas are still in the background, and, further, that it is not clear that Parsons's own line of approach is as close to Cantor's as one might wish. To begin with, then, how can one reconcile Parsons's own characterization of set theory as a very paradigm of platonism with his conception of the 'universe', itself, of all sets (a proper class) as essentially potential and, moreover, akin to intensions seen as, in some sense, constructs of the mind? How can sets themselves be platonic objects if their "totality" is not? Further, has he not leaned, himself, in the very direction of "idealism" he was concerned with on Wang's part, with his own invocation of meanings constructed by the mind? Indeed in this regard, it seems even not quite appropriate to continue to speak of "the universe" of all sets. Is a 'potential-universe' a *universe?* (Recall that Kant's idealism leads him to be suspicious of talk of the [here, physical] universe [as an absolute totality].) Parsons, of course, is well aware of all this, and recognizes that when he chooses to characterize talk of all sets as "systematically ambiguous," this itself threatens the very idea of *the* universe of all sets (296). What is not clear, however, is what has happened, at this point, to set theory as paradigmatic platonism, and to the avoidance of idealism.

A further question that naturally arises for Parsons's modal conception of set existence concerns the 'totality' of the possible worlds themselves. Clearly, he would not accept the notion, introduced earlier in this monograph, of 'modal space'. Indeed, in his 1983d account of modal set theory, he espouses a metatheoretical perspective that operates not with the classical set-theoretic account of possible worlds, but rather with a primitive modal language. What I wish to focus on at this point, however, is the question of the totality of all sets. Parsons moves too quickly, to my mind, from the fact that this multiplicity is "inconsistent" – from the fact, that is, that it does not constitute a *set* – to the conclusion that it is not, strictly speaking, a *totality* at all (which presupposes the existence of all its elements), but rather a mere '*potentiality*'. After all, for Cantor the 'absolute infinity' of an 'inconsistent multiplicity' is not potential but *actual* (Hallett 1984, p. 41). Wang, it seems to me, comes closer to Cantor's original conception. "[T]he multitude V of all sets . . . presupposes for its existence (the existence of) all sets and therefore cannot itself be a set. . . . The universe of all sets is the extension of the concept of set and is a multitude (many) and not a unity (one)" (1977, pp. 310–11). If a set is a 'class as one', then the universe of sets is a 'class as many' (to borrow Russell's terminology from 1903, §70) or a 'manifold'. A manifold, for example, of sheep, is not some *one* thing (viz., a set) that represents a 'collecting-together' of the sheep ('in thought', that is – the collection of sheep in the meadow, of course, is a flock), but rather *many* things – that is, *the sheep!* (See Sharvy, 1980, Simons, 1982, Boolos, 1984.) To be, then, though it is always to be (the same as) *something,* is not always to be *one,* or a '*unity*'. Indeed, as (Plato's) Parmenides says: ". . . the character of unity is one thing, the character of sameness another. This is evident because when a thing becomes 'the same' as something, it does not become 'one'. For instance, if it becomes the same as the many, it must become many, not one." (Plato, *The Parmenides*, 1957, 139d, drawn attention to in Simons, 1982.)

The advantages of viewing the universe of sets as a many, a manifold, are several. For one thing, one need not give up the idea of this as a genuine universe, a totality, that is not merely potential, and that does indeed presuppose the existence of all its members, for nothing prevents the members of a manifold from enjoying ontological parity, that is, from having, all of them, the same ontological status – a characteristic we attributed, earlier, to a 'space', in the general sense. This would agree, then, with Gödel's idea of sets as 'spacelike' (extending the meaning we attributed earlier to this phrase, from the 'quasi-spatial' being of all the members of a set to the quasi-spatial being of all the sets themselves). In addition, we would avoid the troubling question of how one can conceive of a multiplicity as potential without invoking, implicitly, temporal (A-theoretic) notions. It is not clear, that is, how the collection of sets, on Parsons's reading, can be merely *potential,* if it is not becoming *successively completed* (over time). For what is needed seems to be something in between being merely a limited and extendable domain and being a fully completed, unextendable, 'platonistic' domain; and this in between mode of being has traditionally been viewed as that of successive coming-into-being – that is, the very mode of being of time and 'activity' that Parsons was trying to steer us away from in providing an intuitive basis for set theory.

Further, we would avoid, by invoking manifolds, essential use of Cantor's dangerous idea that the members of an "inconsistent multiplicity" cannot "exist together." This phrase seems to be temporal/spatial in origin (thus evoking ideas out of place in set theory), and I remain uncomfortable with what seems to be its modal interpretation by Parsons (existing together in a possible world). To my mind, either we have here something tautologous (the members of a non-setlike multiplicity cannot all exist-together-*in-a-set*), or I am inclined to protest, as Dedekind did to Cantor, that "I don't know what you mean either by the 'being together of all elements of a multiplicity' or its opposite," [letter to Cantor, Aug. 3, 1899, quoted in Hallett (1984), p. 34]. Finally, this conception of a manifold would also enable us to conceive of the 'totality' of possible worlds as a spacelike many. This idea, however, raises a new problem. If, as we suggested, the members of a manifold of possible worlds enjoy ontological parity, this still leaves us with the question: What exactly is this ontological status they all equally enjoy? From our earlier discussion of Lewis, it should be clear that this will not be existence. We will come, shortly, to the issue of just what it will be. We turn first, however, to another, crucial, employment of modal reasoning by Parsons, a consideration of which will enable us to draw a number of strands together.

Parsons puts forward a modal, as opposed to a temporal (A-theoretic), conception of potential infinity. He suggests (1983a, p. 47), as a representation of potential infinity, a model for a modal logic containing S4, in which the possible worlds are all finite, but where for each world there is always one possible relative to it which contains one more member. And in 1983d he continues, in this vein, to suggest that for "conversion of a 'potential' infinity into an 'actual' infinity . . . in terms of a set-theoretic semantics for the modal language, the possible worlds containing finite segments of ω need to be collected into a single one" (pp. 295–6). Now, it should be immediately clear how far this modal conception of the potential/actual infinity distinction is from the temporal (A-theoretic) one of Aristotle, Kant, and Brouwer (if our reconstruction of the ideas of these thinkers has been along the right lines). But let us consider, explicitly, some of the questions that arise for the modal conception of potential infinity. One peculiarity that comes immediately to mind is that we would seem to have before us, now, two different 'sizes' of potential infinity. On the one hand we have the potential infinity of the series of possible worlds representing all the finite segments of ω, and on the other the (presumably) potential infinity of the (essentially) potential multiplicity of (the 'universe' of) all sets. Here, we must face the prospect that although it is difficult to see how one is to compare two different, merely potential infinities, the first is intuitively the 'smaller' of the two – it 'precedes', as it were, Cantor's ascending sequence of actual infinities – while the second not only cannot be "converted into" an actual infinity, but it seems to 'transcend' the entire hierarchy.

A further problem is that, from a certain point of view, the first model is precisely one of an *actual* infinity (even before the "conversion"), for does it not present us with an infinite sequence of possible worlds – that is, an *actually infinite* sequence (if it is really to represent *all* the finite segments of ω)? [It is becoming even clearer why Parsons suggested (1983d) that strictly we should give up the set-theoretic possible worlds approach to modal logic (to fulfill the program of modal set theory).] To deny this implication, it would seem necessary to view the sequence of possible worlds as itself a merely potential multiplicity. But this would seem to bring with it a doubling of modalities (that is, a *potential* totality of *possible* worlds) – and what could this mean? [One must be careful not to conflate what is at issue here with the fact that in classical semantics of modal logic, as in Kripke (1963), we have a clear representation of iterated modal operators, e.g., 'possibly possibly p' (or $\Diamond \Diamond$p). Here, $\Diamond \Diamond$p is true in a possible world α, just in case $\Diamond$p is true in a possible world α', accessible to α, and $\Diamond$p, in turn, is true in α', just in case p is true in a possible world α'', accessible to α'. We thus have clear truth conditions for *formulas,* like $\Diamond \Diamond$p, without needing to invoke such things as 'possible–possible-worlds'.]

Indeed, to connect this issue with our earlier discussions of temporal series, we observe that Parsons's modal model for potential infinity employs, as it were, only 'B-theoretic' (to speak figuratively), 'structural–geometric', features of a manifold, as opposed to the A-theoretic (to speak literally) properties of potential infinity of Aristotle, Kant, and Brouwer. Parsons's model is B-theoretic, or relative-A-theoretic, insofar as the salient feature that we are to read as limiting the infinity in the model to the potential is that we are only permitted to assess the size of the segments of ω present in the model *relative to,* or from the perspective of, individual possible worlds. He does, in fact, say elsewhere that "the alternate possibilities that our modal operators envisage are simply those in which possibilities of set formation have been realized that as yet have not been realized. The logic will then be somewhat like the sort of tense logic in which everything past is necessary (unalterable by later events)" (1983d, p. 316). But this quasi-temporal talk of "realization" is only figurative. Clearly, nothing is really 'happening' in his model: All is, as it were, 'static, geometric, relational'. Our response to the model clearly violates, then, the way we are intended to read it, namely, from the perspective of the individual possible worlds (hence, the significance, in "converting" a potential into an actual infinity, of "collecting together" all finite segments of ω into a single possible world). [Recall our similar problems with the Thomason model of indeterminist time, which, it turned out (or so we claimed) was indeterminist only if one always limits one's 'perspective' to individual times.]

Once again, it would seem, we have a disagreement with Parsons over what is required in order to have before us a truly spacelike multiplicity. As before, Parsons appears to require that all the elements be 'collected together' in a single possible world (where they can all 'exist together', or all 'be actual' [relative to that world]), while on our account the presence of all the members in a manifold – that is, as a many – suffices for us to be presented with a 'space' (here, a 'modal-space'), in which the members enjoy ontological parity. But now, since the manifold in question must span possible worlds, the ontological status shared by all the worlds (and, hence, by all the finite segments of ω) must be, not existence or actuality, but rather possibility.

But what business do we have quantifying over an (actually infinite) domain of mere possibilities? Must there not, after all, be something there to quantify over? Precisely, but that is why we hinted, in our earlier discussion of Lewis and Stalnaker, that we must distinguish *being* (something) from *existing* (being actual, being present), and why we will insist, in the following section, that the existential quantifier, "$\exists x$" or "there is something *x*", be read not, as with Quine (1969, p. 97), as expressing existence, but rather, '*being*' – that is, being something that (if a contingent being, like us) can come to, as well as cease to, exist. Indeed, it is perhaps worthwhile to point out that this conception of (possible) objects (including possible worlds) should not be viewed as *Meinongian*. Parsons, unfortunately, in characterizing the free variables in (one of) his modal constructions of set theory as ranging over all possible objects, expresses concern lest this aspect of his approach be thought "too Meinongian" (1983c, p. 286, note 25). And Kripke, it will be recalled, already got into trouble on just this point, in his seminal (1963), when he incautiously asserted that although Sherlock Holmes does not exist, he might have. Such a claim really does appear to invoke a naive Meinongian semantics that all proper names, including those in fiction, refer to genuine objects. At issue here, however, in letting one's variables range over all possible objects is not a Meinongian semantic thesis about *names,* but a 'Parmenidean' ontological proposal about *objects* (as I put the matter in Yourgrau 1987, p. 91). Thus, as Kripke wrote in his "Addendum" to 1963: "I could no longer write, 'Holmes does not exist, but in other states of affairs, *he* would have existed.' . . . This change affects my view of the linguistic status of fictional names in ordinary language but does not affect the model-theoretic issue involved in the text, namely (1) some of the entities which actually exist might not have existed, and there might have been entities other than those which actually exist . . ." (p. 172).

If, then, distinctions like these are sound, I remain skeptical that the modal approach to potential infinity can succeed in capturing the intuitions that, I have argued, lie behind the classic, temporal (A-theoretic), conception (unless, that is, implicit temporality is smuggled into the proposed model). Thus, here, as with the modal approach to set theory, one finds yet more evidence of the difficulty of doing justice to (what one takes to be) a genuinely non-spacelike, essentially incomplete, 'totality' while restricting one's formal methods to the 'geometric–spatial' (in the general sense). Indeed, I think Parsons's work, here, can be seen, in part, as a strong response to just this dilemma. What our worries have centered on is the question of whether one can succeed, as would appear to be Parsons's intention, in developing non-'geometric–spatial' formal methods that yet "replace the language of time and activity by the more bloodless language of potentiality and actuality." The price of abandoning (A-theoretic) temporality may be a high one, indeed. Things become even worse, however, once we turn from mathematical to more general, ontological concerns, where (A-theoretic) time, it turns out, appears to be essential to marking the distinction between human-being and human-existence.

BEING AND EXISTENCE

It is an unpleasant but unavoidable deliverance of intuitive time that, since we are not gods, there lies in our future the consequence of our mortality – namely, death. ["The coming of death is the inescapable result of the irreversible flow of time. . . . The fear of death is thus transformed into a fear of time" (Reichenbach 1956, p. 4).] This is, I would suggest, one of the phenomena that any philosophy that hopes to be taken seriously must, in Plato's terms, "save." And when we are dead, what then? Do we, somehow, become nothing? Hardly, since the very real grief of our survivors turns on their ability to still think of or refer to us (without thereby thinking of nothing). But who exactly will my survivors be grieving over? Well, me, except myself *as* dead, gone, nonexistent. For past individuals have not merely changed their residence, having simply 'moved into the past'. [Although this is how

Harry Silverstein (1980), following Quine and the (alleged) spatialization of time in the STR, chooses to view things.] Rather, as Aristotle puts it, "[f]or living things, it is living that is existing" (*De Anima* 1983, 415b12).

But if, for us, existing means living, then not living – that is, being dead – means not existing. Thus, we have it that (1) The dead don't exist. When I announced this result in "The Dead" (Yourgrau 1987b), it seems to have produced somewhat of a shock. Not everyone, however, seems to be shocked by this proposition. Kaplan, for example, writes in "Afterthoughts" (1989b) that "[p]ast individuals are also, in my view, nonexistent." In any case, as we have seen, it follows by exactly one logical step from Aristotle's reasonable observation, which is itself a straightforward deliverance of the intuitive conception of time and of living-in-time (and what other kind of living is there for those of us who are not gods?). But let me shock you even further. It follows immediately from (1) that (2) The dead cannot be buried. For what does not exist is obviously not there to be shoveled under the ground. You will certainly bury *something,* only it will not be *me.* This is a point that continues to elude the Critos of this world: "'But how shall we bury you?' said Crito. 'Any way you like,' replied Socrates, 'that is, if you can catch me and I don't slip through your fingers.' . . . 'I can't persuade Crito that I am this Socrates here who is talking to you now . . . ; he thinks that I am the one whom he will see presently lying dead . . . [but] when I have drunk the poison I shall remain with you no longer'" (Plato, *The Phaedo*, 1960, 114A–115D).

Once we have extended our ontological horizons, however, to recognize the dead as nonexistent objects, we are free to acknowledge other nonexistents: the unborn (i.e., future *people*), those who will, alas, never be born (merely possible *people*), merely possible *worlds,* past and future *times,* and so on. Thus, we propose an ontological symmetry here in terms of *categories of being:* Future people are *people* (not *concepts, names,* etc.), past people are *people* (not *memories,* etc.), possible worlds are *worlds* (not sets of *concepts* – Stalnaker 1976 – sets of *sentences,* etc.). And we thus satisfy Kant's famous dictum that "the real contains no more than the merely possible. A hundred real dollars do not contain the least coin more than a hundred possible dollars" (1965, A599 B627). We do not, however, thereby commit ourselves to a spatialization of these categories of being – in the manner of Lewis's extreme and literal "modal realism," since we continue to distinguish the *mode of being* of these beings, namely, *existence* (of present or actual people, times, worlds), from *nonexistence.* Thus, along the lines of our discussion of the metaphysics of modal logic in the previous chapter, we continue to steer a course between the Scylla of Lewis (spatialization) and the Charybdis of Stalnaker (ontological asymmetry).

We thus *take individuals seriously.* Physics is not ontology. The logic of (human) *beings* is not the logic of *bodies.* We are not bodies but are embodied. (As we saw, in our Frege reconstructed, the sentence is not the thought, but rather its embodiment.) What we are in need of, then, to understand human existence is not more physics or biology, but rather a systematic logic of embodiment. [For some important first steps in this direction from an Aristotelian perspective, see P. F. Strawson (1963). For more advanced Aristotelianism, see Montgomery Furth (1988). For Platonism, my own sympathies lie with Harold Cherniss (1962).] Even today, however, and even among modal logicians and metaphysicians, individuals are not, in general, taken fully seriously. In addition to the dilemmas of Lewis and Stalnaker, already addressed, one can point to Kripke, in many respects (as we said), the seminal figure in modal metaphysics. For although Kripke correctly insists, in (1980), as against Lewis (and Leibniz? See Ishiguro 1972, p. 123), on "transworld identity" – namely, that the selfsame *individuals* exist in different possible worlds – when it becomes a question of the ontological status of the possible worlds themselves, he backs off. Fearing, it would seem, to sail his ship toward the Lewis-Scylla or the Stalnaker-Charybdis, he declines to set sail at all. As we saw earlier, we are merely warned, with Lewis in view, not to take the model of possible worlds too literally, but we are not told just *how* literally we *ought* to take them.

To be blunt, it too often appears that the modal logician (or the tense logician) is inclined to back away from the perilous metaphysics toward which his or her mathematical models would appear to point. And when the charge is: *metaphysics!*, the reply is: *mathematics!* But this is both surprising and unfortunate, since, it seems, part of the great interest, for example, in Kripke's and Thomason's formal models in these areas is due precisely to their suggestive *metaphysical* implications. On this matter, Kaplan appears to be more willing to face the metaphysical consequences, with respect to individuals, to which his own work would appear to point. As we have just seen, he recognizes that past individuals do not exist and, reversing his course from "Transworld Heir Lines" (1979), he now recognizes, in "Demonstratives" (1989a), the transworld identity of individuals. But he goes further and correctly points out that since his position requires that individuals not be purely qualitatively individuated, it follows, "although I am uncomfortable with [it] . . . that the world might be in a state qualitatively exactly as it is, but with a permutation of individuals" (p. 505). I take this willingness to follow the idea of the irreducibility of individuals even to such, at first glance, strange conclusions to be a sign that one is now engaged in serious metaphysics.

Adopting, then, an ontological, as opposed to a merely biological, point of view, we pose the question: Where do we come from? (And were are we going?) That babies come from the womb is a familiar biological fact concerning our early *bodies,* but it is of no help with the present question. I thus lay down my third alarming proposition: (3) We don't really know *where* we (as babies) came from. And now something equally surprising begins to emerge. Gödel himself appears to be joining us on this Platonic journey. For in four long letters to his mother in 1961, he lays down some consequences, on the nature of the soul, of his "theological worldview" – "the idea that the world and everything in it has meaning and reason, and in particular a good and indubitable meaning" (Wang 1987, p. 217). (Of course, one must bear in mind here that these letters were not intended for publication.) To begin with, "[i]t follows immediately that our worldly existence, since it has in itself at most a very dubious meaning, can only be means to the end of another existence" (ibid., p. 217). Indeed, in connection with my proposition (3) Gödel writes that "[s]ince we came into existence one day in this world *without knowing how-so and whence,* the same can happen again in the same way in another world" (ibid., p. 214; my emphasis). This is also clearly an echo of the *Phaedo,* where Socrates suggests that the correct picture is not 'living → death' but rather 'death' (i.e., prenatal nonexistence) → living → death. But this last picture, he argues, leads, by unavoidable symmetry, to the most complete account: 'death' → living → death → 'living.' (I put scare quotation marks around the outermost terms, since these represent at best analogues of the middle terms.) Thus, as Socrates says, "the living have come from the dead no less than the dead from the living" (Plato 1960, 71A–72B). For Plato then, as, it would seem, for Gödel, birth and death are *changes* – namely, the crossing of the border between nonexistence and existence. Plato makes this even more explicit in the *Parmenides:* "The non-existent one cannot move by shifting its position . . . and the non-existent cannot *be* in anything that exists" (1957, 162c). (Thus, the *unborn* nonexistent cannot 'be' in the *womb* or the *dead* nonexistent in the *grave.*) But "transition is motion," and "the nonexistent one has been shown to be a thing that moves, since it admits transition from being [nonexistent] to not-being [nonexistent]" (ibid., 162b; my interpolation).

Birth and death, then, are changes (with respect to existence), and as Plato makes clear in *The Phaedo* and as Aristotle affirms in a number of places, change requires a continuing substratum that underlies this 'motion'. Thus, it is the selfsame individual, Socrates, before, during, and after life. Hence, we conclude that Socrates is not to be identified merely with the existing (hence, living) Socrates. And so we have our next thesis, (4): (The) 'Being' (of Socrates) ≠ (the) Existence (of Socrates). Since there must 'be' Socrates before birth in order for his birth to constitute a change in Socrates, from non-existence to existence, being Socrates cannot be the same thing as Socrates existent. Since intuitively, then, there is someone, Socrates, who did not exist but came to exist, I suggested in (1987b) that we let the formal existential quantifier express being – that is, express the fact that there is such a person as Socrates (before and after birth) – and also that we introduce a new, primitive, predicate for existence – namely, "E!". The statement, then, that Socrates is (now) dead – that is, nonexistent – continues to imply, by the principle of existential generalization, that there is someone who is nonexistent. This is now parsed as: $(\exists x)(\sim E!x)$. And Kaplan seems to agree with this when he writes, in "Afterthoughts," that "[i]t would then be natural to add a *narrow existence predicate* to distinguish the robust being of true local existents, like you and me, from the more attenuated being of the nonexistents" (1989b, p. 608). Here, too, however, it is Plato who has set the stage: "[A one which does not exist] . . . must in some sense even possess being. For it must be in the state we are ascribing to it. Otherwise, we should not be speaking the truth in saying that the one does not exist" (1957, 161e) [my interpolation].

But even having distinguished being from existence, and having recognized birth and death as motions or changes of the selfsame being from the state of nonexistence to existence (or the reverse), we have not yet given the complete picture. For living, as we reminded ourselves earlier, is a process that takes place in (clearly, A-theoretic) time, and we have yet to stress the fact that birth obviously *precedes* death. And that is, of course, why we naturally *fear* death but not our prenatal nonexistence – a fact that must remain invisible or inexplicable to spatializers of time like Silverstein (1980) and Mellor (1981) (and recall here our discussion of Mellor's attempt to account for our different attitudes to present vs. future pains and misfortunes). Putting to work our earlier distinction between the order and the direction of a series, then, we note that it is the A-series direction of our deaths as lying in our futures that accounts for the natural fear of dying, since from the point of view merely of B-theoretical temporal order, the positions occupied by pre- and postnatal nonexistence are symmetric. As with the direction of a potentially infinite mathematical series, the A-series direction, here, of living, pointed toward our deaths, is not reversible. And that is why even if all ends for us at death, it matters a great deal, as both Plato and Nietzsche insist, *how* we die, for the end is more important than the beginning. Indeed, in *The Phaedo,* Plato goes so far as to describe philosophy itself as the proper preparation for, and even cultivation of, dying.

Since Gödel, however, seems to be in agreement here with Plato concerning the logic of individual human existence, with reference to our entrance (through embodiment) into and our exit from this world, represented by birth and death, it follows that Gödel's views on Einstein and time, in regard to this world, give us only half a picture of his philosophy. The theological worldview, it seems, must complement the space–time perspective. Once again we are called upon to view the world from two different points of view – from within and also *sub specie aeternitatis.* Yet, somehow, we, the individual selves, must be able to support *both* perspectives. Then what *are* we? "We do not know *what* we are (namely, in essence and seen eternally)" (Gödel in Wang 1987, p. 215). But we do know *that* we can raise such questions and can choose to view this world (including our present embodiment in this world) from a higher (for Gödel, "theological") perspective. In *this* sense, then, we know that "[t]he mind is eternal in so far as it conceives things under the form of eternity" (Spinoza, 1955, *Ethics,* Book V, Proposition XXXI). Perhaps, then, we should consider Gödel's relativistic geometrization of space and time in the light of Plato's own cosmological geometrization in *The Timaeus* (in terms of fundamental geometrical Forms). For it appears at first glance that there can be, for Plato, no place for us in the phenomenal world except as merely one of those flickering images (in the Receptacle of Becoming) of the Fundamental Forms, and that our mode of being – namely, (temporal) becoming – is only a moving *image* of Eternity. Yet Plato recognizes *human-being* as somehow intermediate between merely temporal-phenomenal and genuinely Formal or Ideal being. Thus, as Cherniss (1962) puts it, for Plato, "both ideal being and phenomenal becoming are apprehended by a subject which is neither the one nor the other, but a mean between the two" (p. 410).

Now no one, I believe, from Plato, to Aristotle (recall his proposal that the form of the mind is formlessness), to Kant (the phenomenal vs. the transcendental ego), to Wittgenstein ("The subject does not belong to the world": 1974, 5.632), to Gödel, has succeeded in solving, or "dissolving," this question of the intermediate existence of the self – that self that is somehow able to adopt both perspectives toward the world. But we can at least place Gödel, I think, with Plato, in discerning here a genuine – indeed, a fundamental – clue to the nature of human being that is not dissolved by a more perspicuous view of language or resolved by a synthesis of the two perspectives. For Gödel, with Plato, views our embodied existence, in this world, as at best a confused and cloudy preparation for something more meaningful – in our postnatal nonexistence (i.e., nonexistence after death in this world). "Without a next life," he writes, "the potential of each person and the preparations of this life *make no sense*" (Wang 1987, p. 214), for "[w]hat would be the *point* of bringing forth an essence (the human being) that has so wide a range of possible (individual) developments . . . but is never allowed to realize one thousandth of them?" (ibid., p. 214). And perhaps the following remarks of Gödel on the non-reducibility of mind to machine point in the same direction of unlimited potential: "*[M]ind, in its use, is not static but constantly developing* . . . [and indeed] [t]here is no reason why the number of mind's states should not converge to infinity in the course of its development" (ibid., 197; my interpolation). Further, with Plato in *The Phaedo,* he insists that "[t]he greater part of learning will take place in the next world" (ibid., p. 214) and, moreover, that "we could very well be born [into the next world] with latent memories [of our experiences in this world]" (ibid., p. 214; my interpolations).

It is not a question then, for Plato or for Gödel, of simply living longer, beyond death. Rather, as we saw earlier, eternity means for Plato (and, I would suggest, for Gödel) not "long-lasting life" but rather "genuine" ("Ideal") being. Thus, there is no force here to Wittgenstein's protest that "[n]ot only is there no guarantee of the temporal immortality of the human soul, that is to say of its eternal survival after death; but, in any case, this assumption completely fails to accomplish the purpose for which it has always been intended. Or is some riddle solved by my surviving forever?" (1974, 6.4312). In the first place, as Gödel suggests, it is rather *death* that closes off the unlimited possibilities of individual human development, death that introduces meaninglessness or the riddle of embodied existence. In the second place, it is a question not just of further existence but of development in a certain direction – the direction of (self) perfection, and so transfiguration.

The picture, I believe, is as follows: Since we are beings who are born with the greatest possible separation between what we are potentially and what we are actually (since neither God nor angel has room for improvement), it follows that the purpose of our lives consists not in perfecting any given stage of development, but rather in transcending it, of moving to the next, or higher, stage. As a perfect child is precisely a child successfully on the way to becoming an adult (and not, therefore, a kind of everlasting Peter Pan), a perfect adult points toward something beyond – something that has the same relation to adulthood as the latter has to childhood. ["Man is called a baby by God, even as a child by a man" (Heraclitus, Fragment 97, in Burnet 1968, p. 140).] Naturally, in both cases, the child in us clings to the stage at hand, fearing the darkness of the only dimly discernible future transfiguration, in which we – as we have come to know (and love) ourselves – will suffer the 'death' of unrecognizability.

Having emphasized, however, that, notwithstanding Wittgenstein's suggestions, there is no fallacy lurking behind the Platonic–Gödelian conception of the significance of death, I would like to try to cut a little deeper and to offer a possible means of reconciling what may appear to be two irreconcilable points of view. Gödel and Plato, then, would appear to wish to transcend the limitations of temporal being by contemplating the possibility of crossing the barrier that death places in the way of our embodied existence. Wittgenstein, too, however, considers it essential to escape the confines of temporal being. However, he thinks of accomplishing this from *within* time: "For life in the present there is no death. If by eternity is understood not infinite temporal duration but non-temporality, then it can be said that a man lives eternally if he lives in the present" (1961, p. 75e). [Compare Pascal: "We never keep to the present. . . . Thus we never actually live, but hope to live" (1966, Pensée 47).] What Wittgenstein seems to have in mind here is not the merely negative attitude of ignoring the full temporal context of our lives, but rather just the reverse – that is, rendering the daily fabric of our existence of such significance that each moment approaches the ideal of becoming self-contained or "an end in itself." Far from wishing to neglect "the whole," Wittgenstein wishes each part to reflect it, as a work of art may reflect a whole life:

"The usual way of looking at things sees objects as it were from the midst of them, the view *sub specie aeternitatis* from outside. . . . The work of art is the object seen *sub specie aeternitatis*" (Wittgenstein 1961, p. 83e).

Wittgenstein, I suggest, has in mind, when he speaks of "life in the present," a distinction not unlike that of Aristotle's between *energeia* and *kinesis*. A *kinesis* is an essentially temporal activity that aims at some end external to the process itself (like hunting for your score of the Goldberg Variations), whereas an *energeia* is an activity (or state – see White 1980) that either is an end (*telos*) or in which an end is contained, and that is thus complete, perfect, or self-contained (like contemplating your score of the Goldberg Variations once you've found it). Now the paradigm of an *energeia* is intellectual (aesthetic, loving) contemplation, and this concept, one could argue, has its home in Plato's philosophy. But now Plato seems to be aligned not with Gödel but with Wittgenstein! What has happened? The answer, clearly, is already before us. Gödel and Wittgenstein are at one with Plato that, in essence, temporal being is incomplete being, and that to fulfill our potential most completely as truly human beings we must seek somehow to "cheat" time, to overcome the limitations of temporality. Gödel chooses to emphasize the partial nature of any such representation, or embodiment, of eternity from within the fabric of time (indeed, even Aristotle believes that only God-as-*noûs* could perfectly engage in the perfect *energeia* of contemplation). Wittgenstein, in contrast, wishes to emphasize the possibility of escaping (as much as is humanly possible) the tyranny of death by the incorporation into one's daily life of the perspective of eternity. If Plato can be cited on behalf of each of these attitudes to death and temporality, the reason for this, I believe, is that he himself adopted both.

E. N. Lee (1976) suggests, along these lines, that Plato maintained two points of view on the nature of *noûs* (and so, also, on the nature of the philosophical life and the question of dying). From the "indwelling perspective" (*Timaeus* and *Laws,* Book X), the model of *noûs* is the rotation of a sphere around a fixed and motionless center – an image of pure, perspectiveless cognition unconstrained by the necessity of adopting a limited point of view. [Plato's use of the sphere, here, is surely not unrelated to Parmenides's (1957, p. 44), and Lee remarks, correctly in my view, that for both thinkers this image marks "the real as perspectivally neutral" (p. 98).] In contrast, from the "threshold perspective" (roughly, *Phaedo, Phaedrus, Republic,* and *Symposium*), *noûs* represents the freeing of the *psyche* from its kinetic wanderings in this physical realm of fret and motion, releasing it into the company of the motionless, stable, and perfect Forms, with which it feels a deeper kinship. The guiding metaphor here is that of *noûs* as the culmination of the crossing over from the realm where one finds oneself – but where one is never really at home – to the next (and happier) realm. "It is essentially," says Lee, "the exile's view" (p. 91). But since there is no point in crossing the threshold into the happier realm unless one is prepared to dwell there, and since there is good reason to make our stay or dwelling in our present realm as much like that in the next as possible, it would appear that the two perspectives are complementary. So, too, then, would be the related points of view of Gödel and Wittgenstein.

What is not clear to me, however, is how Gödel can reconcile his Platonic conception of our this-worldly preparation (gradually, over time) for something more meaningful, beyond, with his strong doctrine of the ideality of time, which, though Kantian, is not Kant's, and is clearly other than Plato's. For time, it would seem, for Plato, does not entirely lack being (as it would appear to, for Parmenides – whom Gödel cites approvingly) but rather enjoys a kind of lesser or merely reflected being. Kant, as so often, appears to be a kind of halfway house here, yet it is not clear that Gödel is sympathetic to the idea of the Kantian transcendental self (echoed, to some extent, in the *Tractatus* – a book Gödel is known to have disliked). The resolution of this dilemma, for Gödel, may lie in his well-known affinity for the very different idealism of Leibniz. Yet, to date, we still have the fewest words from Gödel's pen on this, his most admired philosopher. For us, however, there still remains the dilemma of reconciling these two bodies of phenomena – the philosophical concomitants of A-theoretic time and the challenge Gödel has posed to incorporating this intuitive time within the framework of relativity theory. Even more generally, there is the problem throughout of reconciling the *ideal* (the perfect, completed, fully actual) with the *real* (the empirical, merely factual, given). This is, I believe, the fundamental problem in the Platonic philosophy, and reflection on it will enable us to draw a number of strands together, taking our lead from Chapter 5, Section 2 on Frege's decontextualization of thought.

DECONTEXTUALIZATION IN A GENERAL SETTING

What has emerged from our discussion so far is the Platonic point that the context of lived, human, temporal reality suggests or points to an ideal (or Idea, in Plato's sense), in the light of which alone we can *make sense of* this lived reality but that, *if actually realized,* would lead to paradox. For although Plato is famous precisely for having decontextualized, or *separated,* these ideals, or ideas (Vlastos 1987) – and has been criticized on just this point by Aristotle and Kant – I believe that just as according to his philosophy this world is only seen correctly if viewed in the reflected light of the separated Ideas, these Ideas themselves cannot be grasped, or fully made sense of, apart from their human, temporal, dialogical context, in which the need for them first arises. For Plato, of course, only writes philosophy in the form of dialogues, so that we never see the Ideas themselves except as they are seen by us. Indeed, it is *this,* I believe, and not simply a concern with ethics, in the narrow sense, that Plato takes as the Socratic "turn" to the human. In contrast, for the speech of Parmenides, "[t]ranscending human discourse as it does, its expounder transcends humanity and therefore speaks the language not of mortal men but of gods. . . Dialogue, for Plato, is an externalization of the process of thought. . . . [For Plato] genuine dialogue with another is impossible for anyone who will not engage his own soul in dialogue with itself, that dialogue of which the other is only the externalization and embodiment" (Cherniss 1977, p. 20, my interpolation). [This provides, I believe, a hint, which I cannot here follow up, that the 'individualism' that Wang (1987) attributes to Gödel is not incompatible with the dialogical approach favored by Plato. See Yourgrau (1987c, pp. 112–14) and (1989).] With this sense of dialogue or dialectic in mind, then, let us consider several specific contexts where it arises.

a. In geometry, although Plato insists that the idealized, pure, geometric forms are the true subject of the geometers, and even, in *The Timaeus,* that they constitute, in some sense, the very fabric of the material world, they are, paradoxically, themselves invisible. As in the Lewis Carroll story, if all that remains of the Cheshire cat is the pure geometrical crescent of its smile, then the cat will appear to vanish into thin air. Indeed, there is in general a great mystery, to put it mildly, in Plato's conception of the *visible* realm as a reflection and 'imitation' of the *invisible.*

b. Gödel, as we have seen, suggests, with Plato, that our this-worldly, embodied existence makes inadequate sense on its own if it is the final, ultimate, reality about the human – that is, if it remains forever incomplete. Indeed, as I put it elsewhere, "just as in mathematics one 'sees' on the basis of limited but ever increasing iterations that the sequence of natural numbers will, if we adopt the 'arithmetical' point of view, never end, and similarly that two straight lines in Euclidean space drawn in even slight convergence must, 'in principle,' meet, if we permit ourselves to take on a purely geometrical point of view and thus do not confine our attention to the properties of 'straight lines' in physical space, so in human affairs Gödel suggests, in effect, that the 'lines of convergence' from our brief and so imperfect but yet partially beautiful lives can be 'seen' (eventually) to complete themselves, if we do not limit our attention to the natural history of the 'type-human,' but rather, allow ourselves to adopt the 'theological world-view,' which seeks a broader understanding of a type only partially instantiated in the physical world" (1989, p. 407). Yet, paradoxically, as Plato suggests (e.g., in effect, in *The Symposium*), and as Gödel also seems to suggest (Wang 1987, p. 215), the actual *accomplishment* of perfection or completion here would render us kin to the lifeless, *un*thinking Ideas. In a word, the closer we approach the divine Ideas, the more the distinctively *human,* erotic drive for completion dies. [Recall, again, as Plato says in *The Phaedo,* that "true philosophers make dying their profession" (1960, 67[a]).]

c. Sentences, as we saw with Frege, are merely the bodies of thoughts, mere syntactic pictures of that which in essence is nonmaterial – of what is semantic rather than syntactic. Yet a pure Fregean 'ideal' thought, though a necessary condition for the possibility of using sentences to think, seems itself, if taken literally, paradoxical. Recalling the Platonic mystery in a., we note the problem of how we could ever compare a (material) sentence with an (immaterial) thought and learn to associate the one with the other.

d. In the case of mathematics, Platonism postulates the separation of the mathematical Forms, as if they had a life of their own or a separate existence, as a necessary precondition for the fullest comprehension of actual, lived, temporal, forever incomplete and imperfect mathematical practice. This decontextualization of mathematical practice, however, leads, at the cutting edge, to the postulation of structures that keep eluding us, as with the spacelike Cantorian sets – for Cantor's Continuum Hypothesis remains, after the consistency and independence results of Gödel and Paul Cohen, undecided, even after Gödel's attempts to find new axioms. Here, however, the dialectic is enriched by the presence also of a contextualized mathematics, the intuitionism of Brouwer, where the temporal, human element is reintroduced in the very foundations. With this dialectic, however, we come back to our overarching theme concerning the temporal and the real, and so we conclude with some observations about the relevance of contextualized mathematics to the problem of the ideality of time. We close, then, with some thoughts on the question of the temporalization of mathematics.

TEMPORAL MATHEMATICS

It will be recalled that although Plato attempted to banish all traces of time and process from pure, 'formal,' geometry, there remained strong elements of the temporal, at a minimum in arithmetic, in philosophers from Aristotle to Kant. Since Kant, however, there have been two developments in mathematics, pure and applied, pointing in different directions. In pure mathematics, the purification from all traces of the intuitive or the temporal reached a climax toward the end of the nineteenth century when Hilbert produced a formalization of geometry, Frege provided a reconstruction of arithmetic on the basis of his powerful formal, tenseless, quantificational logic (recall the tense-free definition of "successor"), and Cantor provided a set-theoretic foundation for his hierarchy of tenseless, completed, actual infinities.

To be sure, this purification was, as we have seen, not in all cases unproblematic – as in Parsons's suggestion of implicit temporality in certain attempts to provide an intuitive basis for the genetic-iterative conception of set, and as in our own suggestion to similar effect with respect to our attempt to intuitively ground Parsons's own conception of incomplete totalities. Yet, in spite of some possible exceptions, it is clear enough where developments in foundations, in this regard, are heading. Coordinate with this trend, however, is a movement in another direction – the increasing importation of the strictly mathematical into natural science, especially in physics and especially in the relativistic physics of (space–) time – that is, in Einstein's geometrization of the temporal. Thus, whereas the mathematicians, following Plato's lead, were doing their best to take time out of mathematics – or, in general, to decontextualize it – the physicists were busy reintroducing this temporally purified mathematics into (theory of) time. Thus the physicists of time have been employing an instrument designed expressly, by the mathematicians, to be *insensitive* to the temporal! Should we not, however, explore the possibility of a theory of time that employs a formalism less insensitive to its intended domain of application? [Recall the hint from Čapek we alluded to earlier: "Only when succession and change are incorporated into the very foundations of geometry is there a better chance of expressing the dynamic nature of becoming" (1961, p. 35).]

We have already attempted to provide a reconstruction of Frege's semantic framework that would reintroduce the element of context, including and especially temporal context. And we have seen that there are prospects for coping with the threat of temporal determinism, raised first by Aristotle, if, with him, we permit certain modifications in our (meta) logic and semantics. What, then, should we expect from an appropriate mathematical or formal model suitable to the expression of a theory of (A-theoretic) time? From what we have already seen, we can, I believe, set down the following desiderata: (1) Such a mathematics must employ, in its domain of application, incomplete totalities (such as the open future, potentially infinite series, and proper classes). (2) It must strictly disallow the employment of an insufficiently restricted universal quantifier (such as Parmenides's and Wittgenstein's "all that is the case" and Frege's absolute "all"). (3) This mathematics must contain a feature of directionality beyond order (as we have observed in the concept of potential infinity, and more specifically in the Aristotelian–Kantian succession of natural numbers). (4) Its logic, or metalogic, must be a modification or restriction of the classical (as in Aristotle's restriction of bivalence and his modification of the correspondence theory of truth – also advocated, as we saw, by Parsons for such incomplete totalities as proper classes). (5) Complete formalization, as an ideal, must be abandoned, and the construction of limit cases, in the sense introduced in this essay, must be, in principle, ruled out.

Now we have already seen traces of desiderata (1) to (5) in the philosophies of mathematics of Aristotle and Kant, but the approach to foundations in recent times that appears to satisfy in the most striking fashion *all five desiderata* is surely the intuitionism of L. E. J. Brouwer (which, as we have seen earlier, has strong ties with both Aristotle and Kant). Indeed, Brouwer writes explicitly that the First Act of intuitionism "recognizes that intuitionistic mathematics is an essentially languageless activity of the mind having its origin in the perception of a move of time, i.e. of the falling apart of a life moment into two distinct things, one of which gives way to the other, but is retained by memory" (1952, p. 510). He regards the "empty form" of the bare "two-ity" thus born in time to be, when stripped of all quality, what underlies all two-ities, and thereby to be what constitutes the "basic intuition of mathematics." Clearly, this is a temporal conception of mathematics. If we wish, then, to do justice not to the spatial but to the temporal mode of being, is it not more appropriate to turn to Brouwer than to classical, Platonistic mathematics, which is itself the very paradigm of spatiality (in the broad sense)? Of course, the direction of research in the present philosophical climate is clearly toward an increasing formalization of intuitionism, as in the seminal work of Arend Heyting, and some, like W. W. Tait, have argued explicitly that no real work is done by Brouwer's invocation of the unfolding of time. ["These remarks are not intended to refute Brouwer so much as to cast doubt on whether his description of a move of time has any real content" (Tait

1983, p. 174).] It is intuitionism as Brouwer himself conceived it, however, that is my concern here, and I put forward the following suggestion: As Brouwerian intuitionism stands to its formalizations, so does intuitive, A-theoretic time stand to its spacelike projection into indeterministic, branching, tense logic, or, perhaps onto the world lines of Einstein–Minkowski space–time.

If Brouwer's is a temporal mathematics, however, is its temporal basis A- or B-theoretic? Surely, there can be no question but that a B-theoretic intuitionism is an impossibility. That intuitionism satisfies all five desiderata just given for an adequate mathematics of intuitive, A-theoretic time should help put the matter beyond doubt, if we make the case explicit, as follows. In the first place, Brouwer recognizes only incompletable, potential infinity, which is distinctive of the A-series. All B-theoretic structures, I have argued, are complete, and thus an infinity that has its basis in the B-series must be actual. This leads us, in turn, to the next fundamental point: that Brouwer clearly has in mind the successive unfolding of intuitive time, and, as has been argued at length in the preceding pages, there is no genuine succession of times in the B-series. The fact that Brouwer does not explicitly invoke the now, or refer to the past, present, and future, is thus not to the point. [It might be well to recall here that Gödel himself is absolutely clear on just this question: "The existence of an objective lapse of time, however, means (or, at least, is equivalent to the fact) that reality consists of an infinity of layers of the 'now' which come into existence successively" (1949, p. 262).]

Added to these last considerations is the fact that there is an irreversible "directionality," in the A-theoretic sense, in the temporal basis of intuitionism, in which a closed and determinate past gives way to a free, open, and not fully determinate future (reminiscent of Aristotle's model of time). Brouwer speaks, for example, of the Second Act of Intuitionism, "which recognizes the possibility of generating *new* mathematical entities: firstly in the form of infinitely proceeding sequences $p_1, p_2, \ldots$, whose terms are chosen more or less freely from mathematical entities *previously acquired*" (1952, p. 511; my emphasis). Moreover, these free choices may be subject to restrictions that "at any stage may be made to depend on possible future mathematical experiences of the creating subject" (p. 511). Further, "in intuitionist mathematics a mathematical entity is not necessarily *predeterminate,* and may, in its state of free growth, at some time acquire a property which it did not possess before" (1955, p. 552; my emphasis).

Again, along these same lines, what about predictions about what will take place at future times? As Aristotle had before him, Brouwer appeals, in effect, to the A-theoretic intuition that since the future is not 'there' to be referred to, the coming-into-being in the future of a mathematical state of affairs (say, a proof) does not suffice to bring it about that a prediction of the possibility of such a proof in the past was true then. Rather, "expected experiences . . . are true only as anticipations and hypotheses; in their contents there is no truth" (1948, p. 488). Again, the mere linguistic representations of possible mathematical thoughts do not "convey truths *before* these truths have been experienced" (p. 488). [Whether or not, then, one agrees with Dummett that "it would be possible for a constructivist to agree with a platonist that a mathematical statement, if true, is timelessly true: when a statement is proved, then it is shown thereby to have been true all along" (1978b, p. 19), it is clear, I think, that this was not Brouwer's view.] This point of view, however, is derived only in part from the fact that, as with the A-theorist, Brouwer recognizes (in his case, mathematical) reality as itself coming into being successively. In addition, unlike Frege, he refuses to de-contextualize mathematical thought. That is, he resists the crucial step by which Frege, the Platonist, separates, cleanly, thought content from the activity or experience of thinking, in order to then subject this truth-valuable content to the abstract manipulations of a formal logic. Brouwer, of course, is critical of classical logic – especially of the principle of the excluded middle (which, as we have seen, so vexed Aristotle). But the crucial point here is not that Brouwer proposed a revisionist logic. Rather, for him, mathematical truth cannot, without being destroyed, be torn away from its context – the experienced act of cognition. By Brouwer's lights, intuitionism recognized that "the criterion of truth or falsehood of a mathematical assertion was confined to mathematical activity itself, without appeal either to logic or to a hypothetical omniscient being. An immediate consequence was that in mathematics no truths could be recognized which had not been experienced" (1955, p. 552). (For a nice discussion of this point, see Detlefsen 1990b.)

If, then, these suggestions are along the right lines, it would seem that the prospects for an adequate account of a temporalized mathematics like Brouwerian intuitionism cannot be divorced from hopes for a successful elucidation of intuitive, A-theoretic time. Their fortunes, as far as I can see, rise or fall together. It remains, however, that for a mathematical space, like that of Cantor's universe of sets, or for a "rigid four-dimensional space" (to use Einstein's phrase) like Einstein–Minkowski space–time, a more platonistic philosophy is indicated. Perhaps, then, the two perspectives, the intuitionistic and the Platonistic, can be seen to be, in some sense, complementary. Paul Bernays seems to have something like this in mind in his discussion of Gödel's ideas: "Perhaps the methodological analogy [of Gödel's] between foundations of mathematics and of theoretical physics may go even farther, in this way, that just as in the new physical theories the rôle of the observer is an essential ingredient, for the interpretation of mathematics the rôle of mental activity cannot be fully abstracted from. . . . Also, then, the objective, theoretical (existential) standpoint on the one hand and the intuitionistic (constructive) one on the other hand could be regarded as 'complementary' aspects" (1946, pp. 78–9; my interpolation). This still leaves us, however, with the unresolved problem of reconciling the two perspectives for the case of time, "that mysterious and seemingly self-contradictory being which, on the other hand, seems to form the basis of the world's and our own existence," as Gödel (1949) puts it. For, as has become clear, I hope, from the preceding pages, time seems at once to demand and to resist a reconciliation between its subjective, intuitive, manifestations and its appearance in the best models that have been offered to date to incorporate its peculiar being into the fabric of the objective world. It would seem, then, that in constructing the mathematical model of the Gödel universe, Gödel has once again succeeded in raising a traditional, informal question to a new level of clarity, waking us, thereby, from our dogmatic slumbers. For what Frege observed about number can fairly, I think, be said about time: "The first prerequisite for learning anything is thus utterly lacking – I mean, the knowledge that we do not know" (1980, p. III).

References

Almog, J., et al., eds., 1989. *Themes from Kaplan.* New York: Oxford University Press.

Aristotle, 1979. *Categories and De Interpretatione,* trans. J. L. Ackrill. Oxford: The Clarendon Press.

1983. *De Anima,* trans. D. W. Hamlyn. Oxford: The Clarendon Press.

Bell, J. S., 1989. How to teach special relativity, in J. S. Bell, *Speakable and Unspeakable in Quantum Mechanics.* Cambridge: Cambridge University Press, pp. 67–80.

Benacerraf, P., 1965. What numbers could not be, in Benacerraf and Putnam, eds. (1983), pp. 272–94.

and Putnam, H., eds. 1983. *Readings in the Philosophy of Mathematics.* Cambridge: Cambridge University Press.

1985. Skolem and the skeptic, *Proceedings of the Aristotelian Society,* suppl. LIX: 85–115.

Bernays, P., 1946. Review essay: Kurt Gödel, "Russell's mathematical logic," *Journal of Symbolic Logic,* 11, no. 3: 75–9.

Boolos, G., 1984. To be is to be a value of a variable (or to be some values of some variables), *The Journal of Philosophy,* 81: 430–49.

Brouwer, L. E. J., 1948. Consciousness, philosophy, and mathematics, in Heyting (1975), pp. 480–94.

1952. Historical background, principles, and methods of intuitionism, in Heyting (1975), pp. 508–15.

1955. The effect of intuitionism on classical Algebra of logic, in Heyting (1975), pp. 551–4.

Burgess, J., 1978. The unreal future, *Theoria,* XLIV: 157–74.

1979. Logic and time, *Journal of Symbolic Logic,* 44: 566–82.

Burnet, J., 1968. *Early Greek Philosophy.* Cleveland: Meridian Books, World Publishing Company.

Čapek, M., 1961. *The Philosophical Impact of Contemporary Physics.* New York: Van Nostrand Reinhold Company.

Cherniss, H., 1962. *Aristotle's Criticism of Plato and the Academy.* New York: Russell and Russell.

1977. *Selected Papers,* ed. L. Tarán. Leiden: E. J. Brill.

Chihara, C., 1982. A Gödelian thesis regarding mathematical objects: Do they exist? And can we perceive them? *Philosophical Review,* XCI: 211–22.

Church, A., 1951. A formulation of the logic of sense and denotation, in P. Henle, et. al., eds., *Structure, Method, and Meaning: Essays in Honor of Henry M. Sheffer.* New York: Liberal Arts Press, pp. 3–24.

1956. *Introduction to Mathematical Logic.* Princeton, N.J.: Princeton University Press.

1973. Outline of a revised formulation of the logic of sense and denotation, Pt. I, *Nous,* 7: 24–33; Pt. II, *Nous,* 8: 135–56.

Craig, W. L., 1979. The infinite past regained: A reply to Whitrow, *British Journal for the Philosophy of Science,* 30:161–72.

Dauben, J. W., 1979. *Georg Cantor: His Mathematics and Philosophy of the Infinite.* Cambridge, Mass.: Harvard University Press.

Davidson, D., 1968. On saying that, in *Inquiries into Truth and Interpretation.* Oxford: Clarendon Press, 1984, pp. 93–108.

Detlefson, M., 1979. On interpreting Gödel's second theorem, *Journal of Philosophical Logic,* 8, 3: 297–313.

1986. *Hilbert's Program: An Essay on Mathematical Instrumentalism.* Dordrecht: D. Reidel.

1990a. On an alleged refutation of Hilbert's program using Gödel's First Incompleteness Theorem, *Journal of Philosophical Logic,* 19: 343–77.

1990b. Brouwerian intuitionism, *Mind,* 99: 501–34.

Dipert, R. (Weldon, R., coauthor), 1976–7. Set-theoretical music analysis. *Journal of Aesthetics and Art Criticism,* 35:15–22.

Dipert, R., 1982. Set theoretical representations of ordered pairs and their adequacy for the logic of relations, *Canadian Journal of Philosophy,* XII: 353–74.

Dummett, M., 1960. A defense of McTaggart's proof for the unreality of time, in Dummett (1978a), pp. 351–7.

1967. Truth, in Strawson, ed. (1967).

1969. The reality of the past, in Dummett (1978a), pp. 358–74.

1973. The philosophical basis of intuitionistic logic, in Dummett (1978a), pp. 215–470.

1978a. *Truth and Other Enigmas.* Cambridge, Mass.: Harvard University Press.

1978b. *Elements of Intuitionism.* Oxford: The Clarendon Press.

Earman, J., 1967. On going backward in time, *Philosophy of Science,* 34: 211–22.

Eddington, A., 1958. *The Nature of the Physical World.* Ann Arbor: University of Michigan Press.

Einstein, A., 1949. Remarks concerning the essays brought together in this cooperative volume, in Schilpp (1982), pp. 687–8.

1961. *Relativity: The Special and General Theory*. New York: Crown Publishers.

Evans, G., 1985. Understanding demonstratives, in Yourgrau, ed. (1990), pp. 71–96.

Feferman, S., 1984. Kurt Gödel: Conviction and caution, *Philosphia Naturalis,* 21: 546–62.

Frege, G., 1892. On sense and meaning, in P. Geach and M. Black, eds., *Translations from the Philosophical Writings of Gottlob Frege,* 3rd ed. Totowa, N.J.: Rowman and Littlefield, pp. 56–78. (Formerly translated as "On Sense and Reference.")

1967a. The thought: A logical inquiry, in Strawson, ed. (1967), pp. 17–38.

1967b. *Grundgesetze* (*The Basic Laws of Arithmetic*), trans. M. Furth. Berkeley and Los Angeles: University of California Press.

1970. Begriffschrift: A Formula Language Modeled upon That of Arithmetic, for Pure Thought, in J. van Heijenoort, ed., *Frege and Gödel: Two Fundamental Texts in Mathematical Logic.* Cambridge, Mass.: Harvard University Press, pp. 5–82.

1979. *Posthumous Writings,* ed. H. Hermes et al. Chicago: University of Chicago Press.

1980. *Grundlagen* (*The Foundations of Arithmetic*), trans. J. L. Austin. Evanston, Ill.: Northwestern University Press.

Friedman, M. 1983a. Form and content, in Yourgrau (1990), pp. 215–31.

1983b. *Foundations of Space–Time Theories.* Princeton, N.J.: Princeton University Press.

1985. Kant's theory of geometry, *Philosophical Review,* XCIV: 455–506.

1990. Kant on concepts and intuitions in the mathematical sciences, *Synthese,* 84: 213–57.

Furth, M., 1988. *Substance, Form, and Psyche: An Aristotelian Metaphysics.* Cambridge: Cambridge University Press.

Gillies, D., 1988. *Frege, Dedekind, and Peano on the Foundations of Arithmetic.* Assen, The Netherlands: Van Gorcum.

Gödel, K., 1930. The completeness of the axioms of the functional calculus of logic, in Gödel (1986), pp. 103–23.

1931. On formally undecidable propositions of *Principia Mathematica* and related systems I, in Gödel (1986), pp. 145–95.

1944. Russell's mathematical logic, in Gödel (1990), pp. 119–41.

1949. A remark about the relationship between relativity theory and idealistic philosophy, in Yourgrau, ed. (1990), pp. 261–5.

1964. What is Cantor's continuum problem?, in Gödel (1990), pp. 254–70.

1986, 1990. *Collected Works,* Vols. I and II, ed. S. Feferman et al. New York: Oxford University Press.

unpublished. Some observations about the relationship between theory of relativity and Kantian philosophy.

Goldfarb, W. D., 1979. Logic in the twenties: The nature of the quantifier, *The Journal of Symbolic Logic,* 44: 351–68.

Hacker, P. M. S., 1972. Frege and the private language argument, *Idealistic Studies,* 2: 265–81.

Hallett, M., 1984. *Cantorian Set Theory and Limitation of Size.* Oxford: The Clarendon Press.

Heyting, A., ed., 1975. *L. E. J. Brouwer: Collected Works,* Vol. I. Amsterdam: North-Holland.

Hilbert, D., 1902. *The Foundations of Geometry,* trans. E. J. Townsend. Chicago: Open Court.

Horwich, P., 1987. *Asymmetries in Time.* Cambridge, Mass.: MIT Press.

Huby, P., 1971. Kant or Cantor? That the universe, if real, must be finite in both space and time, *Philosophy,* 46: 121–32.

Ishiguro, H., 1972. *Leibniz's Philosophy of Logic and Language.* London: Duckworth.

Jeffrey, R., 1980. Coming true, in *Intention and Intentionality.* C. Diamond and J. Teichman, eds., Ithaca, N.Y.: Cornell University Press, pp. 251–60.

Kant, I., eds., 1902. *Gesammelte Schriften,* Akademie Edition. Berlin: De Gruyter.

1950. *Prolegomena,* trans. P. Carus (revised, L. Beck). Indianapolis: Bobbs-Merrill.

1965. *Critique of Pure Reason,* trans. N. K. Smith. New York: St. Martin's Press.

Kaplan, D., 1978. Dthat, in Yourgrau (1990).

1979. Transworld heir lines, in M. J. Loux, ed., *The Possible and the Actual.* Ithaca, N.Y.: Cornell University Press.

1989a. Demonstratives, in Almog (1989), pp. 481–563.

1989b. Afterthoughts, in Almog (1989), pp. 565–614.

1990. Thoughts on demonstratives, in Yourgrau (1990), pp. 34–49.

Katz, J. J., 1988. The refutation of indeterminacy, *Journal of Philosophy,* LXXXV: 227–52.

Kitcher, P., 1978. The plight of the Platonist, *Nous,* 12:119–36.

Kleene, S. S. *Introduction to Metamathematics.* Amsterdam: North-Holland, 1962.

Kosman, A., 1969. Aristotle's definition of motion, *Phronesis,* 14: 40–62.

Kreisel, G., 1980. Kurt Gödel, *Biographical Memoirs of Fellows of the Royal Society,* 26: 149–224.

Kripke, S., 1963. Semantical considerations on modal logic, in L. Linsky, ed., *Reference and Modality,* Oxford Readings in Philosophy. Oxford: Oxford University Press.

1980. *Naming and Necessity.* Cambridge, Mass.: Harvard University Press.

Lachterman, D., 1989. *The Ethics of Geometry.* New York: Routledge.

Lakatos, I., 1987. *Mathematics, Science, and Epistemology,* Philosophical Papers, Vol. II, ed. J. Worrall and G. Currie. Cambridge: Cambridge University Press.

Lear, J., 1980. Aristotelian infinity, *Proceedings of the Aristotelian Society,* 29: 187–210.

1982. Aristotle's philosophy of mathematics, *Philosophical Review,* XCI: 161–92.

Lee, E. N., 1976. Reason and rotation: Circular movement as the model of mind (noûs) in later Plato, in W. H. Werkmeister, ed., *Facets of Plato's Philosophy.* Assen: Van Gorcum, pp. 70–102.

Lewis, D., 1973. *Counterfactuals.* Cambridge, Mass.: Harvard University Press.

Malament, D., 1984. "Time-Travel" in the Gödel-universe, *Proceedings of the Philosophy of Science Association,* 2: 91–100.

McTaggart, J. M. E., 1908. The unreality of time, *Mind,* 17: pp. 457–84.

1927. *The Nature of Existence,* Vol. 2. Cambridge: Cambridge University Press.

Mellor, D. H., 1981. *Real Time.* Cambridge: Cambridge University Press.

Mohanty, J. N., 1984. Husserl, Frege, and the overcoming of psychologism, in K. K. Cho, ed., *Philosophy and Science in Phenomenological Perspective.* Dordrecht: Martinus Nijhoff, pp. 143–52.

Moore, W., 1989. *Schrödinger: Life and Thought.* Cambridge: Cambridge University Press.

Moravcsik, J. M. E., 1973. Recollecting the theory of forms, in Werkmeister, ed. (1976), pp. 1–20.

Murdoch, I., 1978. *The Fire and the Sun: Why Plato Banished the Artists.* New York: Oxford University Press.

Myhill, J., 1951. On the ontological significance of the Löwenheim–Skolem theorem, in M. White, ed., *Academic Freedom, Logic, and Religion.* Philadelphia: University of Pennsylvania Press, pp. 57–70.

Newton-Smith, W., 1980. *The Structure of Time.* London: Routledge & Kegan Paul.

Nietzsche, F., 1968. *Twilight of the Idols,* trans. R. J. Hollingdale. Baltimore, Md.: Penguin.

Parmenides, 1957. *Way of Truth,* trans. F. M. Cornford, in *Plato and Parmenides.* Indianapolis: Library of Liberal Arts, Bobbs-Merrill.

Parsons, C., 1968. Review essay: Jose A. Benardete, *Infinity, Journal of Philosophy,* LXV: 431–7.

1980. Mathematical intuition, in Yourgrau (1990), pp. 195–214.

1983. *Mathematics in Philosophy.* Ithaca, N.Y.: Cornell University Press.

1983a. Ontology and mathematics, in Parsons (1983), pp. 37–62.

1983b. The liar paradox, in C. Parsons (1983), pp. 221–67.

1983c. What is the iterative concept of set? in Parsons (1983), pp. 268–97.

1983d. Sets and modality, in Parsons (1983), pp. 298–344.

Pascal, B., 1966. *Pensées,* trans. A. J. Krailsheimer. London: Penguin Books.

Penrose, R., 1989. *The Emperor's New Mind.* Oxford: Oxford University Press.

Perry, J., 1977. Frege on demonstratives, in Yourgrau (1990), pp. 50–70.

Plato, 1957. *The Parmenides,* trans. F. M. Cornford. Indianapolis: Bobbs Merrill Co.

1960. *The Phaedo,* in *The Last Days of Socrates,* trans. H. Tredennick. Harmondsworth, Middlesex, England: Penguin Books.

1971. *Timaeus and Critias,* trans. D. Lee. Harmondsworth, Middlesex, England: Penguin Books.

1973. *The Sophist,* trans. F. M. Cornford, in *Plato: Collected Dialogues,* ed. E. Hamilton and H. Cairns, Bollingen Series LXXI. Princeton, N.J.: Princeton University Press.

1973a. *Theaetetus,* trans. J. McDowell. Oxford: Clarendon Press.

Popper, K., 1982. *The Open Universe.* Totowa, N.J.: Rowman and Littlefield.

1984. A critical note on the greatest days of quantum theory, in A. O. Barut, et al., eds., *Quantum, Space and Time – The Quest Continues.* Cambridge: Cambridge University Press, pp. 49–54.

Prior, A. N., 1959. Thank goodness that's over, *Philosophy,* 34: 12–17.

1967. Time and determinism, in *Past, Present and Future.* New York: Oxford University Press.

1970. The notion of the present, *Studium Generale,* 23: 245–8.

Putnam, H., 1979. Time and physical geometry, in H. Putnam, *Mathematics, Matter, and Method,* Philosophical Papers, Vol. I. Cambridge: Cambridge University Press, pp. 198–205.

ed., 1983. *Readings in the Philosophy of Mathematics,* Benacerraf (1983).

1985. Models and reality, in H. Putnam, *Realism and Reason,* Philosophical Papers, Vol. 3. Cambridge: Cambridge University Press, pp. 1–25.

Quine, W. V. 1969. Existence and quantification, in W. V. Quine, *Ontological Relativity and Other Essays*. New York: Columbia University Press.

1979. *Word and Object*. Cambridge, Mass.: MIT Press.

Reichenbach, H., 1956. *The Direction of Time*. Berkeley, Calif.: University of California Press.

1958. *The Philosophy of Space and Time*. New York: Dover Press.

Rietdijk, C. W., 1966. A rigorous proof of determinism derived from the special theory of relativity, *Philosophy of Science*, 33: 341–4.

Rosen, S., 1961. Thought and touch: A note on Aristotle's *De Anima*, in Yourgrau, ed. (1990), pp. 185–94.

1988. Heidegger's interpretation of Plato, in *The Quarrel Between Philosophy and Poetry*. New York: Routledge & Kegan Paul, pp. 127–47.

Russell, B., 1903. *The Principles of Mathematics*. Cambridge: Cambridge University Press (2nd ed., New York: Norton, no date)

1915. On the experience of time. *Monist*, 25: 212–33.

1956. Mathematics and the metaphysicians, in J. R. Newman, ed., *The World of Mathematics*, Vol. III. New York: Simon & Schuster, pp. 1576–90.

Schilpp, P. A., ed., 1963. *The Philosophy of Rudolph Carnap*. La Salle, Ill.: Open Court Press.

1982. *Albert Einstein: Philosopher-Scientist*. La Salle, Ill.: Open Court Press.

Sharvy, R. 1980. A more general theory of definite descriptions, *The Philosophical Review*, 89: 607–24.

Silverstein, H., 1980. The evil of death, *Journal of Philosophy*, LXXVII: 401–23.

Simons, P. 1982. II Number and manifolds, III Plural reference and set theory, in Simons, Three Essays in Formal Ontology, in B. Smith, ed., *Parts and Moments: Studies in Logic and Formal Ontology*. München: Philosophia Verlag, pp. 160–260.

Sklar, L., 1974. *Space, Time, and Space–Time*. Berkeley: University of California Press.

1981. Time, reality and relativity, in Yourgrau, ed. (1990), pp. 247–60.

1984. Comments on Malament's "'Time-Travel' in the Gödel-Universe," *Proceedings of the Philosophy of Science Association*, 2: 106–10.

1985. *Philosophy and Spacetime Physics*. Berkeley: University of California Press.

Sorabji, R., 1986. Closed space and closed time, in M. Woods, ed., *Oxford Studies in Ancient Philosophy*, Vol. IV. Oxford: Clarendon Press, pp. 215–31.

Spinoza, B., 1955. *On the Improvement of the Understanding, The Ethics,*

Correspondence, trans. R. H. M. Elwes. New York: Dover Publications, Inc.

Stalnaker, R., 1976. Possible worlds, *Nous,* 10: 65–75.

Stein, H., 1968. On Einstein–Minkowski space–time, *The Journal of Philosophy,* 65: 5–23.

1970. On the paradoxical time-structures of Kurt Gödel, *Philosophy of Science,* 37: 589–601.

1990. Introductory note to K. Gödel (1949), in Gödel (1990), pp. 199–201.

Strawson, P. F., 1963. *Individuals.* Garden City, N.Y.: Anchor Books.

ed., 1967. *Philosophical Logic,* Oxford Readings in Philosophy. Oxford: Oxford University Press.

1985. *The Bounds of Sense:* An Essay on Kant's *Critique of Pure Reason.* London: Methuen.

Tait, W. W., 1983. Against intuitionism: Constructive mathematics is part of classical mathematics, *Journal of Philosophical Logic,* 12: 173–95.

Thom, R., 1971. "Modern" mathematics: An educational and philosophic error?, in T. Tymoczko, ed., *New Directions in the Philosophy of Mathematics.* Boston: Birkhäuser, 1985, pp. 67–78.

Thomason, R., 1970. Indeterminist time and truth-value gaps, *Theoria,* XXXVI: 264–81.

Tiles, M., 1989a. Philosophy and the analogies of time, *The Philosophical Forum,* XX: 182–94.

1989b. *The Philosophy of Set Theory: An Historical Introduction to Cantor's Paradise.* Oxford: Basil Blackwell.

Tragesser, R., 1984. *Husserl and Realism in Logic and Mathematics.* Cambridge: Cambridge University Press.

Van Fraassen, B., 1985. *An Introduction to the Philosophy of Time and Space.* New York: Columbia University Press.

Vlastos, G. 1987. 'Separation' in Plato, in J. Annas, ed., *Oxford Studies in Ancient Philosophy,* V. Oxford: Clarendon Press.

Wang, H., 1961. Process and existence in mathematics, in Y. Bar-Hillel et al., eds., *Essays on the Foundations of Mathematics, Dedicated to A. A. Fraenkel on His Seventieth Anniversary.* Jerusalem: Magnes Press, Hebrew University, pp. 328–51.

1974. *From Mathematics to Philosophy.* London: Routledge & Kegan Paul.

1977. Large sets, in R. E. Butts and J. Hintikka, eds., *Logic, Foundations of Mathematics and Computability Theory.* Dordrecht: D. Reidel, pp. 309–33.

1987. *Reflections on Kurt Gödel.* Cambridge, Mass.: MIT Press.

forthcoming. To and from philosophy, *Synthese.*

Webb, J., 1980. *Mechanism, Mentalism, and Meta-Mathematics.* Dordrecht: D. Reidel.

Werkmeister, K., ed., 1976. *Facets of Plato's Philosophy, Phronesis,* Suppl. Vol. 2: 1–20.

Weyl, H., 1949. *Philosophy of Mathematics and Natural Science.* London: Princeton University Press.

White, M. J., 1980a. Necessity and unactualized possibilities in Aristotle, *Philosophical Studies,* 38: 287–98.

1980b. Aristotle's concept of Θεωρία and the 'Ενέργεια-Κίνησις distinction, *Journal of the History of Philosophy,* 18: 253–63.

1981. Fatalism and causal determinism: An Aristotelian essay, *The Philosophical Quarterly,* 31: 231–41.

1985. *Agency and Integrality.* Dordrecht: D. Reidel.

1989. Aristotle on "Time" and "A Time," *Apeiron,* XXII: 207–24.

Whitrow, G. J., 1978. On the impossibility of an infinite past, *British Journal for the Philosophy of Science,* 29: 39–45.

1980. *The Natural Philosophy of Time.* Oxford: The Clarendon Press.

Wilson, M., 1984. The "Phenomenalisms" of Berkeley and Kant, in A. Wood, ed., *Self and Nature in Kant's Philosophy.* Ithaca, N.Y.: Cornell University Press, pp. 157–73.

Wittgenstein, L., 1961. *Notebooks 1914–1916,* G. H. von Wright and G. E. M. Anscombe, eds., trans. Anscombe. New York: Harper and Brothers.

1974. *Tractatus Logico-Philosophicus,* trans. D. F. Pears and B. F. McGuinness. London: Routledge & Kegan Paul.

1978. *Remarks on the Foundations of Mathematics,* ed. G. H. von Wright et al., 3rd ed. Oxford: Basil Blackwell.

Yourgrau, P., 1982. Frege, Perry, and demonstratives, *The Canadian Journal of Philosophy,* 12: 725–52.

1983. Knowledge and relevant alternatives, *Synthese,* 55: 175–90.

1985a. Sets, aggregates, and numbers, *The Canadian Journal of Philosophy,* 15, no. 4: 581–92.

1985b. Russell and Kaplan on denoting, *Philosophy and Phenomenological Research,* XLVI: 31–321.

1985c. On the logic of indeterminist time, *The Journal of Philosophy,* 82, no. 10: 548–59.

1986. On time and actuality: The dilemma of privileged position, *The British Journal for the Philosophy of Science,* 37: 405–17.

1987a. Frege on truth and reference, *Notre Dame Journal of Formal Logic,* 28: 132–8.

1987b. The dead, *The Journal of Philosophy,* 84, no. 2: 84–101.

1987c. The path back to Frege, in Yourgrau, ed. (1990), pp. 97–132.
1989. Review essay: Hao Wang, *Reflections on Kurt Gödel, Philosophy and Phenomenological Research,* 50, no. 2: 391–408.
ed., 1990. *Demonstratives,* Oxford Readings in Philosophy. Oxford: Oxford University Press.

Index